Manufacturing technology

Manufacturing technology for level-2 technicians

Bruce J. Black, C. Eng., M.I. Prod. E.
Senior lecturer in industrial engineering,
Gwent College of Higher Education

Edward Arnold

To my late grandfather, David Duncan J.P.

© Bruce J. Black 1983

First published 1983
by Edward Arnold (Publishers) Ltd
41 Bedford Square, London WC1B 3DQ

Edward Arnold (Australia) Pty Ltd
80 Waverley Road
Caulfield East 3145
PO Box 234
Melbourne

All Rights Reserved. No part of this publication may be reproduced, stored in a retrieval system, or transmitted in any form or by any means, electronic, mechanical, photocopying, recording or otherwise, without the prior permission of Edward Arnold (Publishers) Ltd.

British Library Cataloguing in Publication data

Black, Bruce J
   Manufacturing technology for level-2 technicians.
   1. Production engineering
   I. Title
   670.42   TS176

   ISBN 0-7131-3485-2

Filmset in 10/11pt English Times by Colset Private Ltd, Singapore, and printed in Great Britain by Richard Clay (The Chaucer Press) Ltd, Bungay, Suffolk.

Contents

Preface vii

Acknowledgements viii

1 **Welding processes** 1
   Oxy–acetylene welding. Filler metals and fluxes. Oxy–acetylene cutting. Manual metal arc welding. Joint preparation. Positions of welding. Weld defects. Bend testing. Safety in arc welding. Safety in oxy–acetylene welding.

2 **Primary forming processes** 26
   Forms of supply of raw materials. Properties of raw materials. Sand casting. Rolling. Extrusion. Drawing. Forging. Selection of a primary process. Hazards in primary process work.

3 **Presswork** 44
   Presses. Press-tool design. Blanking, piercing, and bending operations. Blanking layouts.

4 **Plastics moulding** 67
   Forms of supply. Moulding processes. Inserts. Advantages and limitations of moulding processes. Safety in plastics moulding.

5 **Measurement** 79
   Length. Straightness. Flatness. Squareness. Roundness. The sine bar. Calibrated steel balls and rollers. Comparative measurement. Sources of error. Methods of comparative measurement.

6 **Single-point metal cutting** 120
   Cutting force. Tool life. Power consumption. Determination of power consumption. Positive and negative rake. Cutting-tool materials. Cutting fluids. Types of cutting fluid. Application of cutting fluids. Safety in the use of cutting fluids.

7 **Machine tools** 146
   Generating, forming, and copying. Power transmission of machine tools. Control of dimension and form. Accuracy of slideway systems. Alignment testing.

8 Operational planning 172

Index 180

Preface

I have written this book to cover the objectives of the Technician Education Council standard unit Manufacturing technology II (U80/736) which is designed to replace U76/056 for technician students of mechanical and production engineering. The aim of the unit is to extend the student's understanding of the basic technological processes required by all mechanical- and production-engineering technicians.

As with my previous books, my aim has been to set out in detail the theoretical aspect of each topic, with appropriate illustrations, in the hope that, by using the book as a course text and for assignments, the student can spend more time in practical work where machines and equipment can be demonstrated, used, and handled or – in the case of equipment not being available – so that time can be effectively spent on works visits, e.g. relating to plastics processing and presswork. Questions directly related to the text are included at the end of each chapter.

Since writing *Manufacturing technology for level-3 technicians*, which was based on the TEC unit Manufacturing technology III (U80/737) but included objectives from earlier units, some of these objectives have now been included in this unit at level 2. Consequently I have included here some material from my level-3 book, in particular welding processes and some material on capstan lathes and machine-tool alignments. I have also felt it appropriate to include cutting-tool materials and cutting fluids from my level-1 book, *Workshop processes, practices, and materials*, especially for those students who received credit for the level-1 unit.

I would once again like to thank my wife and children for their patience, understanding, and assistance throughout the period of writing and preparing the illustrations; my colleagues for their assistance and constructive comments; and Mrs Vanda Hunt for her speedy and skilful typing of the manuscript. Finally, I would like to express my gratitude to Bob Davenport of Edward Arnold (Publishers) Ltd for all his help.

Bruce J. Black

Acknowledgements

The author and publisher would like to thank the following organisations for their kind permission to reproduce photographs or illustrations:

T. Norton & Co. Ltd (fig. 3.1); Sweeney & Blocksidge (Power Presses) Ltd (fig. 3.2); Verson International Ltd (fig. 3.3); P. J. Hare Ltd (fig. 3.4); Lloyd Colley Ltd (fig. 3.13); Tappex Thread Inserts Ltd (figs 4.6 and 4.7); TI Coventry Gauge Ltd (figs 5.2, 5.5, 5.7, and 5.21); Verdict Gauge (Sales) Ltd (figs 5.17, 5.30, and 5.34); J. E. Baty & Co. Ltd (fig. 5.32); Sigma Ltd (figs 5.35 and 5.36); Thomas Mercer Ltd (fig. 5.37); H. W. Ward & Co. (fig. 7.10); Alfred Herbert Ltd (fig. 7.11); Cincinnati Milacron Ltd (figs 7.22 and 7.23); Engineering & Scientific Equipment Ltd (fig. 7.26); Warner Electric Ltd (fig. 7.30).

Figure 4.3 is based on an illustration in ICI technical service note G103, *The principles of injection moulding*, by permission of Imperial Chemical Industries PLC.

Figures 4.4, 4.5, 5.6, 5.10, and 5.22 were photographed by John Kelly.

Table 5.1 is taken from BS 4311:part 1:1968 and Table 5.2 is taken from BS 5317:1976 by kind permission of the British Standards Institution, 2 Park Street, London W1A 2BS, from whom copies of the complete standards may be obtained.

1 Welding processes

General objective The student understands the principal welding processes.

British Standard BS 499:part 1:1965, 'Welding, brazing and thermal cutting glossary', defines a weld as 'a union between pieces of metal at faces rendered plastic or liquid by heat or by pressure or by both'.

There are many different ways in which metal can be welded in accordance with the above definition – fusion processes, where the metal is melted to make the joint with no pressure involved; resistance welding, where both heat and pressure are applied; and pressure only, applied to a rotating part where the heat is developed through friction, as in friction welding.

Table 1.1 shows some of the principal welding processes, of which oxy–acetylene and manual metal arc welding will be discussed in detail in this chapter.

Table 1.1 Welding processes

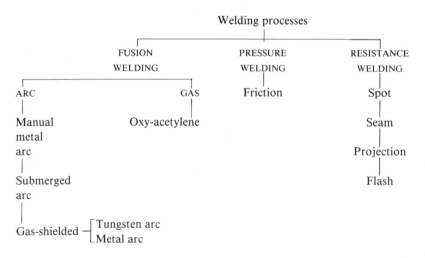

Also included, although not a welding process, is the use of a mixture of oxygen and combustible gas in metal cutting.

1.1 Oxy-acetylene welding

Specific objective   To describe, with the aid of sketches, the principles of oxy-acetylene welding.

Oxy-acetylene welding uses the combustion of the two gases oxygen and acetylene to create a source of intense heat, providing a flame temperature of around 3250°C. This heat is sufficient to melt the surfaces of the metals being joined, which run together and fuse to provide the weld. Additional material in the form of a filler rod may sometimes be required.

The gases are supplied at high pressure in steel cylinders which are made to rigid specifications. Low-pressure systems are also available. Oxygen at a pressure of 172 bar is supplied in cylinders painted black and provided with a right-hand-threaded valve. Acetylene at a pressure of 15 bar is supplied in cylinders painted maroon and provided with a left-hand-threaded valve. The use of opposite-hand threads prevents incorrect connections being made.

Pressure regulators are fitted to the top of each cylinder, to reduce the high pressure to a usable working pressure of between 0.13 and 0.5 bar. The regulators also carry two pressure gauges, one indicating the gas pressure to the torch, the other indicating the pressure of the contents of the cylinder.

The welding torch or blowpipe consists of a body, a gas mixer, an interchangeable copper nozzle, two valves for the control of the oxygen and acetylene, and two connections for their supply. The body serves as a handle so that the operator can hold and direct the flame. The oxygen and acetylene are mixed in the mixing chamber and then pass to the nozzle, where they are ignited to form the flame. The interchangeable nozzles each have a single orifice or hole and are available in a variety of sizes determined by the orifice diameter. As the orifice size increases, greater amounts of the gases pass through and are burned to supply a greater amount of heat. The choice of nozzle size is determined by the thickness of the work, thicker work requiring a greater amount of heat and therefore a larger nozzle size. Welding-torch manufacturers supply charts of recommended sizes for various metal thicknesses and the corresponding gas pressures to be used.

Rubber hose reinforced with canvas is used to supply gas from the regulator to the torch and is colour-coded: blue for oxygen, red for acetylene. The connection for attaching the hose to the torch contains a check valve to prevent the flow of gas back towards the regulator, and it is essential that the hose is correctly fitted.

An illustration of the complete welding outfit is shown in fig. 1.1.

Goggles must be worn at all times when welding. They should be of good quality, provide ample ventilation, fit comfortably, and give proper eye protection. Safety aspects of oxy-acetylene welding are dealt with in greater detail in section 1.10.

The oxy-acetylene welding process can be applied to steels up to and exceeding 25 mm thick, but is mainly used on thinner gauges up to 16

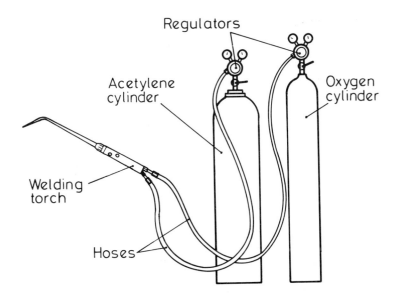

Fig. 1.1 Oxy-acetylene welding outfit

s.w.g. (1.6 mm), where heat input needs to be flexible, and for welding die-castings and brazing aluminium where the heat must be maintained within a critical range.

The oxy-acetylene welding flame

Specific objective   To distinguish between the three types of flame used in oxy-acetylene welding.

In oxy-acetylene welding it is the flame which does the work and is therefore most important. All the welding equipment already described merely serves to maintain and control the flame. It is the operator who sets the controls to produce a flame of the proper size, shape, and condition to operate with maximum efficiency to suit the particular conditions.

The oxy-acetylene torch can be adjusted to produce three distinct types of flame: neutral, carburising, and oxidising.

Neutral flame   When the torch is lit, the acetylene is turned on and ignited, giving a large very smoky yellow flame. The torch acetylene valve is adjusted until the flame ceases to smoke. If the torch oxygen valve is gradually opened until a light-blue inner cone of flame is sharply defined, fig. 1.2(a), at this stage equal quantities of acetylene and oxygen are being used, combustion is complete, and the flame is neutral. This type of flame is the one most extensively used and has the advantage that it adds nothing to the metal being joined and takes nothing away – once the metal has fused, it is chemically the same as before welding.

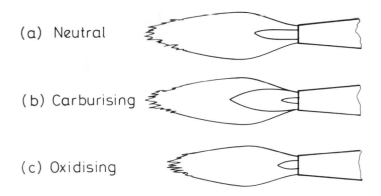

Fig. 1.2 Types of oxy-acetylene flame

The hottest part of the flame is the tip of the inner cone, which, when welding, should be held close to but clear of the molten pool of metal.

Carburising flame This flame is indicated by a feathery white plume around the inner cone, denoting excess acetylene, fig. 1.2(b). The feathery white plume is known as the acetylene feather and varies in length according to the amount of excess acetylene. This excess acetylene is very rich in carbon. When carbon is applied to red-hot or molten metal, it tends to combine with any iron present, forming a hard brittle substance known as iron carbide. The presence of brittle iron carbide can leave the weld unfit for many applications, and the use of a carburising flame for welding is to be avoided.

Use is made of this type of flame, however, when hard-facing – a process in which a layer of hard metal is deposited on the surface of a soft metal to give localised resistance to wear.

Oxidising flame If the flow of acetylene is reduced, the inner cone is shorter, much bluer in colour, and more pointed than the neutral flame, fig. 1.2(c). This indicates excess oxygen, which tends to combine with many metals, especially at high temperatures, to form hard, brittle, low-strength oxides. For this reason, oxidising flames should be avoided in welding.

An oxidising flame is used when welding brass and bronze, as the use of a neutral flame can create porosity, which results in poor mechanical properties.

Methods of oxy-acetylene welding

Specific objective To distinguish between leftward and rightward welding techniques.

Leftward welding This method is used on steel sheet up to 4.5 mm thick and on non-ferrous metals. The weld is started at the right-hand end of the work and progresses towards the left, with the torch held in the right hand. The filler rod precedes the torch and is held at an angle of 30° to 40°, while the torch is held at an angle of 60° to 70° to the work surface, fig. 1.3. This gives an angle of about 90° between filler rod and torch.

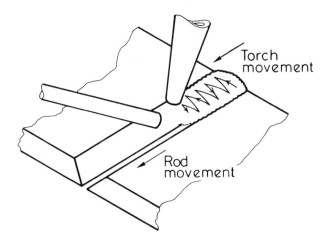

Fig. 1.3 Leftward welding

As welding proceeds, the flame is given a slight side-to-side movement to maintain melting of both edges and obtain even fusion. For butt joints on plate above about 3 mm thick, the edges are chamfered to give an included angle of 80°. This means that more filler metal is used to fill the vee. If the chamfer angle is reduced, however, the flame tends to push the molten metal forward on to the unmelted portion, resulting in poor fusion.

As the thickness of metal being joined increases, a larger nozzle is required, which is more difficult to control; more filler is required, due to the increased size of the vee; and the side-to-side movement over the wider vee makes it difficult to obtain a good even fusion on each side and penetration to the bottom. (Penetration is the depth to which the parent metal, i.e. the metal being joined, has been fused.) As a result, it is necessary to weld thicker plate with two or more layers or 'runs'. The large volume of molten metal present causes considerable expansion, the subsequent contraction leading to distortion.

This process therefore has disadvantages when used on thicker plate, requiring more filler, using more gas, and as a result achieving a slower welding speed. It is recommended that butt welding of steel plate over 5 mm thick is carried out by the rightward welding method.

Rightward welding With this method, welding progresses from the left-hand end of the work towards the right, with the torch held in the right hand. The filler rod in this case follows the torch and is held at an angle of 30° to 40°, while the torch is held at an angle of 40° to 50° to the work surface, fig. 1.4. The torch is moved in a straight line along the weld, while the filler rod is given a circular motion. Using this method, good fusion can be obtained with plate up to 8 mm thick without any chamfer. Above 8 mm, the plate edges are chamfered to an included angle of 60°, the heat being generated in a narrower vee than in leftward welding giving good fusion.

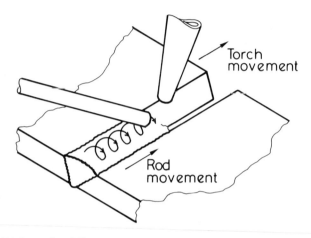

Fig. 1.4 Rightward welding

A larger nozzle is used than with the leftward method, giving a larger flame leading to a faster welding speed. Less filler is used, due to the narrower vee, and the smaller volume of deposit reduces the amount of expansion and results in less distortion. As the flame points backwards towards the part just welded, this helps to slow down the cooling rate, giving an annealing effect.

This method has no advantage over the leftward method on plate thicknesses below 5 mm, but has the following advantages for thicker plate:

a) faster welding speeds,
b) more economical use of filler rod and gas,
c) less distortion,
d) annealing effect on weld metal.

1.2 Filler metals and fluxes

Specific objective To state the type and forms of supply of filler metals and the purpose and forms of supply of fluxes used for oxy–acetylene welding.

Filler metals are chosen to suit, and are generally of the same composition as, the metals being welded.

Most metals in a molten state absorb oxygen from the atmosphere and become oxidised. Fluxes are used to ensure that oxidation is kept to a minimum and that any oxides formed will be dissolved or floated off, which helps make the welding process easier and ensures the making of a good sound weld.

Low-carbon steels The oxide formed when welding these materials has a lower melting point than the parent metal and, being light, floats to the surface as a scale which is easily removed. No flux is therefore required when welding these materials.

Filler metals for use with these materials are low-carbon-steel rods, copper-coated to minimise corrosion in storage. They are available in diameters of 1.6, 2.0, 2.4, 3.2, and 5.0 mm – 750 mm long.

Stainless steels The flux used when welding these materials is a grey powder with a melting point of 910°C which may be mixed with water to form a paste. Its use ensures perfect penetration and freedom from oxides. The residue may be removed by treating in a 5% solution of boiling caustic soda, followed by washing in hot water.

Filler metals for use with these materials are available to suit a number of different types of stainless steel and are available as rods in diameters of 1.6, 2.4, and 3.2 mm – 750 mm long.

Cast iron The oxide formed when welding this material has a higher melting point than that of the parent metal, and it is therefore necessary to use a flux which will combine with the oxide and also protect the metal from oxidation during welding. The flux combines with the oxide and forms a slag which floats to the surface and prevents further oxidation. The flux used is a grey powder with a melting point of 850°C and may be mixed with water to form a paste. The residue may easily be removed by means of a chipping hammer or a wire brush.

Filler metal for use with this material is a high-silicon cast iron and is available as square cast rods 5 mm and 6 mm square, 550 mm long.

Brass and bronze When welding these materials, a flux must be used to remove all oxide from the metal surfaces. The flux must float the oxide and any impurities to the surface and form a protective coating to prevent further oxidation. The flux used is a light-blue powder with a melting point of 875°C and can be mixed with alcohol to form a paste. The residue may be removed by washing in boiling water, followed by brushing.

Filler metal used with these materials is silicon bronze and is available as rods of diameter 1.6, 2.4, 3.2, 5.0, and 6.0 mm – 750 mm long. This type of rod is also available as flux-coated and self-fluxing, eliminating the need to have fluxes on the shop floor.

Aluminium and aluminium alloys A flux must be used when welding these materials, to attack and dissolve the ever-present film of aluminium oxide and to prevent further oxidation during welding. The flux must melt at a lower temperature than the parent metal and be lighter, so that it will float any impurities to the surface where they can be removed.

Aluminium welding flux is a white powder with a melting point of 570°C and may be mixed with water to form a paste. Owing to its corrosive action, care must be taken to remove all traces of the flux after welding, to prevent corrosion. As soon as possible after welding, the work should be washed in warm water and brushed vigorously. When conditions allow, this should be followed by a rapid dip in a 5% solution of nitric acid and washing again, using hot water to assist drying.

Filler metal is available as pure aluminium rods for welding pure aluminium, aluminium-alloy rods containing 5% or 10% silicon for welding all aluminium alloys except those containing magnesium, and aluminium-alloy rods containing 5% magnesium for welding aluminium alloys containing up to 5% magnesium. The rods are available in diameters of 1.6, 2.4, 3.2, and 5.0 mm – 750 mm long.

Copper A flux is recommended when welding this material using copper–silver filler rod and is a white powder with a melting point of 850°C. This flux may be mixed with water to form a paste. The residue may be removed by washing in boiling water, followed by brushing.

Filler metal for use with this material is copper–silver: a low-melting-point phosphorus-deoxidised copper containing a controlled percentage of silver. Rods are available in 3.2 mm diameter, 750 mm long.

1.3 Oxy-acetylene cutting

Specific objective To describe, with the aid of sketches, the principles of oxy-acetylene cutting.

Oxy-acetylene cutting is used to cut through steel and can be used on plate in excess of 300 mm thick. In principle, two operations are involved. First, the steel is heated by means of the oxy-acetylene flame to a temperature around 950°C, known as the oxygen-ignition temperature. A stream of high-pressure pure oxygen is then directed on to the hot metal and causes the metal to ignite and burn to produce more heat. This additional heat causes the nearby metal to burn, so that the process is continuous once it has started. The burning metal is blown away in the oxygen stream.

Only those metals which oxidise rapidly can be flame cut, and these include the plain-carbon steels. If the cutting torch or blowpipe is moved along the material, the metal surface is preheated to a temperature above the ignition temperature and is ignited by the cutting-oxygen jet and a cut is produced. The width of the cut, or 'kerf' as it is known, is governed by the size of the cutting-oxygen jet.

High-pressure cutting equipment is similar to high-pressure oxy-acetylene welding equipment in so far as the gases are used from high-pressure cylinders. Although acetylene fuel gas produces the highest flame temperature, which can be an advantage for speed of reaching the ignition temperature from cold, the use of propane or butane on long cuts may be preferable due to their lower cost. Once the ignition temperature has been reached, the cutting speeds using these gases are often identical, e.g. a cutting speed of 280 mm/min can be achieved with acetylene or propane fuel gas cutting 25 mm thick steel plate.

Gas-cutting torches and nozzles are, however, different from those used in welding. An extra control is required for the cutting oxygen, and this takes the form of a spring-loaded lever located on the torch handle, fig. 1.5. When depressed, this lever releases a stream of pure oxygen through a central hole in the interchangeable nozzle. The diameter of the hole used depends on the thickness of plate being cut.

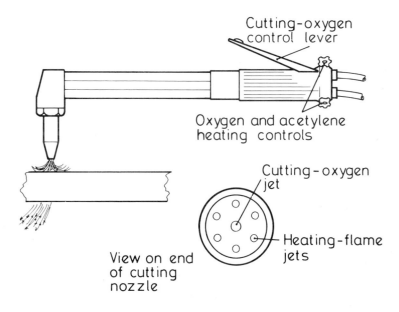

Fig. 1.5 Oxy-acetylene cutting torch and nozzle

Arranged round the central hole are a number of holes (fig. 1.5) at which the mixture of heating gases – in this case oxygen and acetylene – burns to provide the required flame. The flame is adjusted to give the correct mixture of oxygen and acetylene, resulting in a series of short blue luminous cones similar to the neutral welding flame.

Care should be taken to choose the correct size of torch and nozzle to suit the application and metal thickness and to select the appropriate oxygen and acetylene operating pressures. Values for these are supplied by the equipment manufacturers, and typical values are shown in Table 1.2.

To operate, the flame is lit, and the blowpipe is held at right angles to the metal surface and the correct distance from it. The distance of the nozzle from the metal surface depends on the metal thickness and may be, for example, between 3 mm and 5 mm for metal up to 50 mm thick. The edge of the metal furthest from the operator is then heated bright red, i.e. above the oxygen-ignition point, and the cutting-oxygen lever is operated. With cutting taking place, the torch is then drawn towards the operator along a previously marked line, at the recommended speed for the particular conditions. Speeds for set conditions are recommended by the equipment manufacturers, and examples are given for high-pressure torches and acetylene fuel gas in Table 1.2.

Table 1.2 Data for high-pressure oxy–acetylene cutting

Plate thickness (mm)	Nozzle size (mm)	Operating pressures (bar)		Approx. cutting speed (mm/min)
		Oxygen	Acetylene	
6	0.8	1.8	0.14	430
13	1.2	2.1	0.21	360
25	1.6	2.8	0.14	280
50	1.6	3.2	0.14	200
75	1.6	3.5	0.14	200
100	2.0	3.2	0.14	150

High-quality cuts with a regular edge may be difficult to achieve with a hand-held torch. Uniform speeds and a constant distance of the nozzle from the surface are difficult to maintain, resulting in an irregular edge and reduced accuracy. Mechanical aids such as roller or point guides are available and can be used to obtain better results for straight lines and circles than with a solely hand-held torch.

Greater accuracy can be obtained by the use of a cutting machine. One or more cutting heads are positioned above a table upon which the plate to be cut is placed. The cutting heads are mounted on a trolley arrangement which can be moved in two planes. By locking in one direction, the torch or torches can produce a straight-line cut in the other, and vice versa. By unlocking in each direction and moving in both planes simultaneously, in conjunction with a template, an accurate profile of any shape can be cut. The operation of these machines may be manual or automatic.

A quantity of thin plate can be cut as a stack of plates clamped tightly together.

Fire is a particular hazard when flame cutting, since sparks can travel as far as 9 m along the floor. It is essential never to work near combustible or flammable materials. As well as eye protection, suitable protective clothing should be worn, i.e. gauntlets, armlets, aprons, and leggings.

1.4 Manual metal arc welding

Specific objective To describe, with the aid of sketches, the principles of manual metal arc welding (a.c. and d.c.).

This process is the most widely used form of arc welding and is often referred to simply as 'electric arc welding'.

The basic principle is to connect the work to be welded to one terminal of an electrical supply and to connect a metal electrode to the other terminal. These two parts of the electric circuit are brought together and then separated slightly. The electric current jumps the gap and causes a continuous spark known as the arc. The high temperature of this arc – around 4000°C – is sufficient to melt the metal being welded and forms a molten pool. The electrode also melts and adds metal to the pool. As the arc is moved, the metal behind it fuses as it solidifies. The melting action is controlled by changing the electric current flowing across the arc and by changing the size of the electrode. The circuit is shown in fig. 1.6.

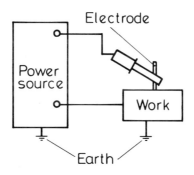

Fig. 1.6 Manual metal arc welding – basic circuit

Either direct or alternating current can be used with manual metal arc welding, and is supplied by means of

a) a transformer, for a.c. welding;
b) a generator or rectifier for d.c. welding.

Transformers The mains electricity supply is not suitable for manual metal arc welding. A transformer is necessary to reduce the high-voltage low-amperage mains supply to a low-voltage high-amperage supply required for arc welding. The output from the transformer is between 80 V and 100 V – known as the open-circuit voltage – with various current capacities from 100 A to 350 A. A higher voltage is required to strike the arc than to maintain it, and so welding sets are designed to automatically drop this voltage to 20 V to 40 V – known as the arc voltage – once the arc has been struck. Due to the rapid change in polarity, heat is developed equally at the work and the electrode (when using direct current, about two thirds of the heat is developed at the positive pole and one third at the negative).

Transformers do not generate electricity and therefore can only be used when a mains supply is available.

Generators Welding generators are used to generate direct current with an open-circuit voltage between 40 V and 60 V. Various sizes are available to produce a range of current capacities.

Generators may be driven by an electric motor if a mains supply is available, or by a diesel or petrol engine which makes them completely mobile.

Rectifiers A rectifier is used to convert alternating current from a transformer to direct current before it is delivered to the circuit. An advantage is that a transformer/rectifier unit can offer both a.c. and d.c. forms of supply.

Advantages and disadvantages of a.c. welding
An a.c. transformer is the most economical type of welding plant in respect of initial, running, and maintenance costs. It has no moving parts and is quiet in operation.

Its disadvantages are that it requires a mains supply and is therefore not mobile. Also, due to the higher open-circuit voltages, the risk of electric shock is greater than with d.c. equipment.

Some electrodes do not perform satisfactorily with a.c.

Advantages and disadvantages of d.c. welding
Practically all electrodes perform well with d.c.

Polarity can be changed to develop the majority of heat at either the work or the electrode, giving greater flexibility. (About two thirds of the heat is developed at the positive pole and one third at the negative using direct current.)

Where no mains supply is available, diesel- or petrol-driven generators are essential.

Direct current is safer than alternating current.

The disadvantages are that generators are noisy and their initial, running, and maintenance costs are higher than for transformers.

D.C. welding is prone to arc blow. ('Arc blow' is the term used when the arc is deflected by the magnetic field in the conductors. It does not occur with a.c. equipment as the effect is cancelled due to the magnetic field constantly changing direction at the frequency of the supply.)

Electrodes
Metallic electrodes are made of various materials, usually to give a deposit of similar composition to the material being welded. They are obtainable in sizes within a range from 2.5 mm to 6.3 mm in diameter and are generally 450 mm long.

The electrodes are coated with a flux, the composition of which varies widely but contains slag- and gas-forming materials, fluxing agents, and alloying elements. The action is shown in fig. 1.7.

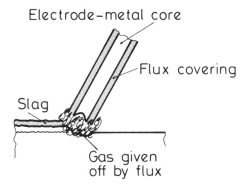

Fig. 1.7 Action of manual metal arc welding

The functions of the flux coating are

a) to protect the molten weld pool from atmospheric contamination by excluding oxygen and nitrogen;
b) to provide a slag over the completed weld to continue to protect the weld from the atmosphere and to control the rate of cooling (the slag has subsequently to be chipped off the completed weld);
c) to stabilise the arc, i.e. keep the arc constant along the length of the weld;
d) to replace alloying elements lost during the welding process.

Manual metal arc welding is applied to steels, especially low-carbon steel, to a greater extent than to other metals. Because of this, a wide range of carbon-steel electrodes is available. Due to the wide variation of welding characteristics, the electrodes are classified and coded in British Standard BS 639:1976, making it easy to identify the different types.

In addition to electrodes with carbon-steel core wire, electrodes are available for welding stainless and heat-resisting steels, cast iron, aluminium, and bronze, and also for hard-facing (i.e. building up on a soft metal, layers of a harder metal capable of withstanding wear).

In storing electrodes, great care should be taken to protect them from damage and from damp conditions. Electrodes with damaged coatings should not be used.

Applications
Manual metal arc welding is used in all spheres of engineering, including shipbuilding, bridge building, and oil and chemical pipelines, and for all types of metal – ferrous and non-ferrous – particularly carbon steels, stainless and alloy steels, and cast iron, 1.6 mm and above.

1.5 Joint preparation

Specific objective To describe, with the aid of sketches, the types of edge preparation necessary for welding.

There are four types of joint common in welding:

i) butt,
ii) tee,
iii) corner,
iv) lap.

Joints (i), (ii), and (iii) can be made by means of a butt weld or a fillet weld. Joint (iv) is used with sheet material and can be made by means of a fillet weld or a spot or seam weld.

In butt welding, the parts being joined are butted together and the joint is made by fusion through the thickness of metal. Fillet welding is carried out using a 'fillet' of weld metal providing local fusion between adjacent faces.

These weld joints and types of weld are shown in fig. 1.8.

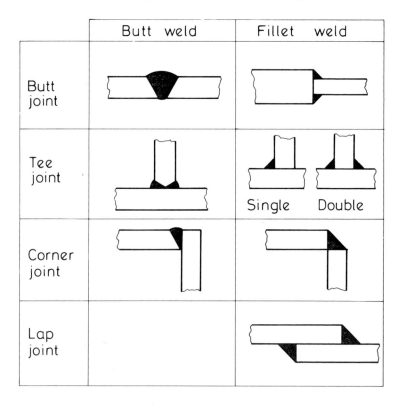

Fig. 1.8 Weld joints and types of weld

In order to achieve the necessary penetration across the thickness in butt welding, edge preparation is usually necessary. The maximum thickness of plate which can be butt welded using square edges (i.e. no edge preparation) is governed by the welding process used, e.g. up to 8 mm using oxy–acetylene rightward welding. When thicker metal has to be joined, the edges must be chamfered to obtain good penetration. The groove so formed is then filled by depositing a number of weld runs fused to each other and to the sides of the groove.

The more usual preparation is the vee groove from one or both sides. As the thickness of metal increases, a single vee groove requires an increased volume of weld metal and the double vee can be used to advantage – though there must, of course, be access from both sides. Use of the double vee balances shrinkage and gives less angular distortion than a single vec. Use of a single 'U' gives an approach to uniform weld width through a section.

Various edge preparations are shown in fig. 1.9.

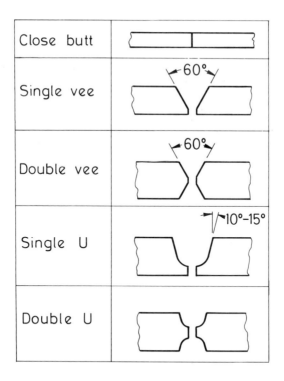

Fig. 1.9 Edge preparation

1.6 Positions of welding

Specific objective   To identify the positions in which welding may be undertaken.

The ideal position for all welding is when the work is level, so that the molten weld metal is held in position by gravity. This is known as the flat position. However, this position is not always possible in all fabrication and construction work, and so difficulties are encountered in achieving proper fusion.

There are three main positions which can be identified in addition to flat; these are horizontal, vertical, and overhead. There is a subdivision of the horizontal position known as horizontal-vertical which relates specifically to tee joints in which the axis of one member is vertical while the other is horizontal. In all these cases the molten weld metal tries to run out of the joint under the action of gravity. This effect is overcome by lowering the heat input and by using a number of passes to build up the weld and maintain proper fusion.

The various positions are shown in fig. 1.10.

1.7 Weld defects

Specific objective   To list visually identifiable defects in welds.

The design of a welded structure is based on the assumption that the weld will have mechanical properties at least equal to those of the parent metal. The presence of defects in the weld may mean that the structure is unacceptable for the purpose for which it is intended.

No welded joint will be absolutely perfect, and what has to be considered is the level of defect which can be tolerated while the joint is still able to satisfy the design requirement. This means that the welded joint has to have a minimum standard in order to withstand particular service conditions.

One way in which this standard can be established is to produce a test-piece of the particular welded joint prepared under the same conditions as those intended for the final welded structure. Small specimens from the test-piece can then be subjected to various tests designed to establish whether or not it is likely to meet the design requirement. What is actually being tested, however, is the ability of the welder or the settings of the welding process.

Obviously such methods, known as destructive testing, are of no practical use in detecting defects in a welded structure ready for service, since the joint would be destroyed and therefore be of no further value. However, a series of tests is available to determine the quality of a welded joint while not damaging the joint in any way. These are known as non-destructive tests, and the simplest of them is visual inspection. It must be appreciated that these tests cannot establish whether or not a weld is acceptable – they are merely used to attempt to detect the type, size, and

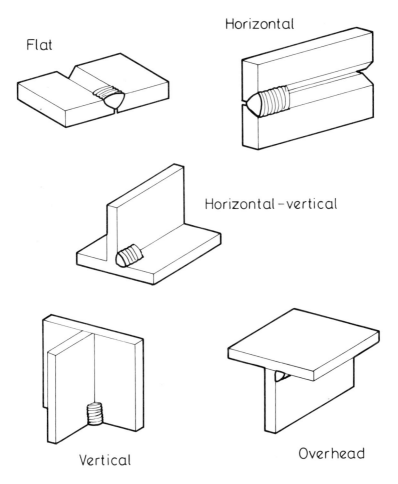

Fig. 1.10 Positions of welding

position of any defects which may be present and so enable a judgement to be made about the likely behaviour of the weld.

Visual inspection is carried out by the naked eye or with the aid of low-magnification lenses. In some cases a dye penetrant is used as a means of attracting the eye to the defect. Dye penetrants may incorporate a fluorescing agent which permits viewing to be carried out under ultra-violet light and enables greater sensitivity to be achieved. Visual inspection can, of course, only be used to discover surface defects and so it is not capable of detecting all of the defects likely to be present in a weld.

The more usual aspects considered in a visual inspection are as follows.

a) *Weld size and profile* The specified size of the weld must be maintained in order to satisfy the design strength; for example, the dimen-

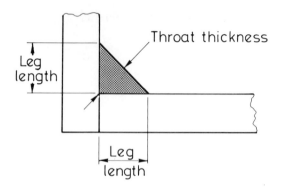

Fig. 1.11 Fillet-weld dimensions

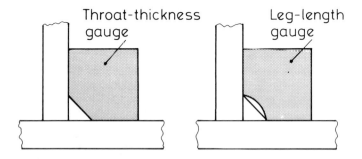

Fig. 1.12 Fillet-weld gauges

sions of fillet welds are indicated by leg length and throat thickness as shown in fig. 1.11. These dimensions can be checked using simple gauges as shown in fig. 1.12. The profile of the weld is also important and will affect the efficiency of the joint; for example, if the surface is concave the throat thickness will be reduced, fig. 1.13. The ripples on the surface should be smooth, rounded, and evenly spaced.

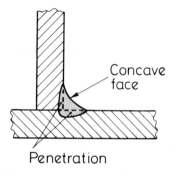

Fig. 1.13 Fillet weld with concave surface

b) *Penetration in joints welded from one side* Penetration is the depth to which the parent metal has been fused and is of the utmost importance in the strength of welds. Single-vee butt welds should show a penetration bead, fig. 1.14, which is even and continuous and soundly fused into the root faces.

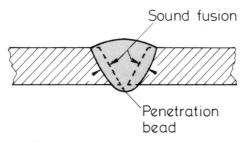

Fig. 1.14 Butt weld showing penetration bead

Fillet welds should have penetration reaching completely into the corner, fig. 1.13. Lack of fusion and of root penetration is not permitted.

c) *Surface defects* A number of defects can appear on the surface of the weld, most of which are detrimental to the weld quality:

   i) *Undercutting* Gravity exerts a force on fluid metal, so there is a tendency for the fluid metal to drop away from the parent metal. The parent metal will be thinner by the amount that drops away, and this thinning of the section is known as undercut, fig. 1.15. Since undercutting reduces the thickness, the parent metal is weakened. In some cases slight intermittent undercuts may be permitted.

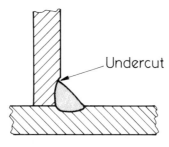

Fig. 1.15 Fillet weld with undercutting

   ii) *Porosity* The presence of a group of small holes or gas pockets in a weld is known as porosity. The holes may be on top of the weld, where they can be seen, or inside the weld. Their presence weakens the weld and in most cases is not permitted.

iii) *Cracks*   When the crater at the end of a weld bead cools, cracks can appear due to hot shrinkage because the crater is thinner than the rest of the hot metal. It is also possible for cracks to appear in the heat-affected zone, i.e. the area around the weld metal which is not melted but is subjected to high temperature. Cracks may also appear if a weld is subjected to too great a load and may be a sign that the weld is too small. The presence of a crack or cracks results in weakness and is therefore not permitted.

iv) *Slag inclusions*   During the melting of metal in fusion-welding processes, oxides are formed. With the correct welding technique and use of fluxes, these oxides can be floated to the top as slag. However, small particles of slag can become trapped in the weld and these are known as slag inclusions. Their presence in a weld is acceptable within certain permitted maxima.

v) *Overlap*   An imperfection at the boundary between the weld face and parent metal caused by metal flowing on to the surface of the parent metal without fusing to it is known as overlap. Such a defect is not permitted.

vi) *Spatter*   Metal particles which do not melt into the molten pool and are thrown off by welding are known as spatter. Spatter on the surface does not generally affect the weld but does indicate faulty techniques.

1.8 Bend testing

Specific objective   To conduct a bend test.

A bend test is a destructive test carried out on a specimen cut from a test-piece and is covered by BS 709:1971. The object of such a test is to determine the soundness of the weld metal, weld junctions, and heat-affected zone (i.e. the area around the weld metal which is not melted but is subjected to high temperature). The test may also be used to give a measure of the ductility of the weld zone.

The test specimen is prepared by dressing the upper and lower surfaces of the weld flush with the original surface of the material. The test specimen is then placed across two roller supports with the centre point of the weld positioned under the centre point of a radiused former as shown in fig. 1.16. The distance between the roller supports is equal to the diameter of the former plus 2.2 times the material thickness of the specimen. The diameter of the former is normally 4 times the material thickness of the specimen. A load is applied to the former, pushing the specimen between the roller supports until its sides are parallel. After bending, the appearance of the joint is examined for flaws, e.g. cracks and tears, to determine its soundness and the ductility of the weld zone.

Normally two specimens are cut from the test piece. One is subjected to a face bend test in which the top surface of the weld is in tension, fig. 1.17(a); the other is subjected to a root bend test in which the root of the weld is in tension, fig. 1.17(b).

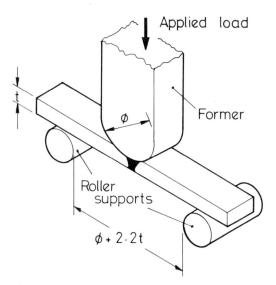

Fig. 1.16 Set-up for bend test

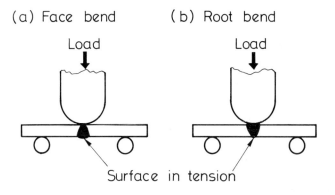

Fig. 1.17 Types of bend test

1.9 Safety in arc welding

Specific objective To state sources of danger in arc welding.

There are a number of dangers related to the use of arc welding equipment with which you should be familiar.

Electric shock
Depending on the conditions, if a human body contacts a live conductor, a current may flow through the body. The current can cause violent

muscular spasms which can cause the body to be flung some distance or fall. In severe cases the heart will stop beating. There is always a risk of electric shock with all electrical equipment, and this is especially true of arc welding, since the welder handles the work and equipment and is therefore particularly at risk.

The risk of electric shock is greater with a.c. welding equipment than with d.c. welding equipment. This is due to the higher open-circuit voltage – between 80 V and 100 V a.c. compared with between 40 V and 60 V d.c. For this reason alone, d.c. is often perferred. Low-voltage safety equipment is available for a.c. equipment which automatically reduces the open-circuit voltage to approximately 30 V when the arc is not being struck.

The welding circuit must be connected using properly insulated cable capable of withstanding the high current used, and all connections should be clean and tight. The circuit consists of three important connections:

i) a welding lead connecting the electrode holder to the power supply,
ii) a welding return connecting the work to the power supply,
iii) an earth lead connecting the work to a suitable earth point.

Electrode holders should be insulated and not have any exposed metal parts. Welding gauntlets will not give sufficient protection from electric shock and should not be relied upon. Welding in damp or wet conditions increases the risk of shock and should be avoided. Wetness or moisture at surfaces increases the possibility of leakage of electricity by lowering the resistance.

In the event of someone suffering electric shock, know what to do:

a) If the victim is still in contact with the electric current, switch off or remove the plug.
b) If the current cannot be switched off, take special care to stand on a dry non-conducting surface and pull the victim clear using a length of dry cloth, a jacket, or some other item of clothing. Remember – *do not* touch the victim as you will complete the circuit and also receive a shock.
c) When free, immediately apply artificial respiration and call for medical assistance.

Placards showing the detailed procedure to be followed in the event of electric shock must be permanently displayed in factories. You should read these carefully and be fully conversant with the procedures – it could save a life.

Radiation from the arc
The arc emits a high level of light radiation, the intensity of which depends on the current and the presence of a flux. Ultra-violet and infra-red radiation are also emitted, which are outside our visible range. All are damaging, and exposure to the eyes is prevented by a coloured filter fitted in the welder's shield. This filter reduces the light to an acceptable level and

absorbs ultra-violet and infra-red radiation. The correct filter is chosen from a recommended range set out in British Standard BS 679:1959 and depends on the current rating and the arc-welding process. A replaceable plain-glass cover is placed over the filter, to protect it from spatter.

Ultra-violet radiation is damaging to both skin and eyes, and even brief flashes of ultra-violet light viewed from a distance can be sufficient to cause a painful eye condition known as 'arc eye'. The symptoms of 'arc eye' are watering eyes, sensitivity to light, a gritty feeling under the eyelids, intense pain, headaches, and in some cases temporary loss of vision. In addition to protection for the welder, adequate screening of the welding position must be provided to protect other workers.

At the onset of symptoms of 'arc eye', protect the eyes from light. To obtain relief, cover the eyes with a cloth frequently wrung out in cold water. The advice of a doctor should be sought if the condition does not clear up quickly.

The skin may also be affected by ultra-violet radiation, slight exposure resulting in a sun-tan effect. Over exposure, however, results in severe burning and blistering. Suitable protective clothing should therefore be worn to prevent any area of skin being exposed.

Burns
Burns may result from sparks or hot metal thrown out by the arc and from the welder picking up or touching hot metal. Burns can be avoided by using the correct protective clothing.

In the event of a burn,

a) put a sterilised dressing on the burn;
b) never use an adhesive dressing and do not apply any lotions, ointments, or oil dressings;
c) do not burst blisters or touch the burned area, thereby increasing the risk of infection.

If the burn is serious, send promptly for a doctor or an ambulance.

Fire
The risk of fire is ever present when arc welding, as molten metal can travel considerable distances. Arc welding should never be carried out in close proximity to flammable substances.

Know the correct fire drill and the positions of fire alarms and fire-fighting equipment.

Know the correct appliance to use for a particular type of fire and how to use it.

Chipping of slag
Where a flux is used, as in the case of manual metal arc welding, the slag formed has to be removed. Goggles must be used to protect the eyes from flying particles during the slag-chipping operation.

The most usual eye injury is particles lodged under the eyelid, causing considerable discomfort and inflammation if not speedily removed. In such a case,

a) prevent the casualty from rubbing the eye;
b) if the particles cannot be speedily removed with sterilised cotton wool moistened with water, close the eyelids, cover them with a soft pad of cotton wool secured lightly in position with a bandage, and obtain medical aid.

All serious injuries to the eye must receive immediate medical aid.

Fumes

Fumes are produced during arc welding, and adequate ventilation is essential. This is particularly important when working in confined spaces. Although most fumes produced are non-toxic, they can cause irritation to the nose and throat. Toxic fumes are given off when welding zinc- and cadmium-plated components and beryllium-copper alloys.

The gases used in gas-shielded arc welding are heavier than air and accumulate at low levels, reducing the oxygen content and increasing the risk of suffocation. In extreme cases, the use of breathing apparatus is necessary.

When working in enclosed vessels, such as boilers and tanks, someone should know you are there and, if possible, someone should be on standby in case of emergency.

In the event of someone being overcome by fumes in an enclosed space, breathe in and out several times and then take a deep breath and hold it before going in to get the casualty out. The use of a lifeline when entering an enclosed space is a safety precaution which should be adopted whenever available. Once rescued, if the casualty's breathing is failing or has stopped, start resuscitation immediately.

1.10 Safety in oxy-acetylene welding

Specific objective   To state sources of danger in oxy-acetylene welding.

Heat for oxy-acetylene welding is generated by combustion, and this can lead to reduced oxygen in the surrounding atmosphere. In confined spaces this can increase the risk of suffocation.

Light and infra-red radiation from an oxy-acetylene flame are intense, and welders must wear goggles with the correct filter as recommended in BS 679.

Acetylene is an unstable gas and at critical concentrations can cause explosions. Great care must be taken to ensure that all connections are tight and that there are no leaks.

Oxygen must not feed back along the acetylene hose and vice versa, since this could lead to burning in the hoses which could spread to the cylinders. This feature – known as flash back – can be prevented by fitting check valves to the torch and flash-back arrestors at each regulator output.

All connections carrying oxygen must be free of grease, as grease contacting oxygen can cause an explosion.

Compressed-gas cylinders should be correctly handled and stored. Always store in well ventilated cool store rooms, with oxygen and combustible gases kept separate. Acetylene cylinders should always be stored and used upright.

Exercises on chapter 1

1 Give two functions of the flux coating on a coated metallic electrode used in manual metal arc welding.
2 What is 'arc blow'? Why does it not occur with a.c. welding?
3 By what name is the cut known in oxy-acetylene cutting?
4 When a destructive test is carried out on a weld, what is being tested?
5 Why are skin and eye protection required against an electric arc?
6 Name the three types of oxy-acetylene flame. How is each produced and for what type of welding is each used?
7 What is 'penetration' and why is it important?
8 Why is it dangerous to weld in an enclosed area such as a tank?
9 State the objectives of a bend test.
10 State the definition of a weld.
11 Why should grease be kept away from pure oxygen?
12 Sketch the end of a nozzle used in flame cutting, and indicate the function of each orifice.
13 What steps should be taken in the event of someone suffering electric shock?
14 Why are pressure regulators necessary on oxygen and acetylene cylinders?
15 Name the two basic types of weld used in fusion welding.
16 What are 'slag inclusions' and why are they caused?
17 What is the purpose of a rectifier?
18 State two advantages of rightward welding over leftward welding for plate thicknesses above 5 mm.
19 Name the four common types of welding joint.
20 What is an 'undercut' and why should it be avoided?

2 Primary forming processes

General objective The student understands the basic primary forming processes.

Most metal objects have at some stage in their manufacture been shaped by pouring molten metal into a mould and allowing it to solidify. On solidifying, the object is known as a casting if the shape is such that no further shaping is required – it may only require machining to produce the finished article. Castings are produced by various methods, for example sand casting, die-casting, and investment casting. If, upon solidifying, the object is to be further shaped by rolling, extrusion, drawing, or forging, it is known as an ingot, pig, slab, billet, or bar – depending on the metal and the subsequent shaping process – and these are cast as simple shapes convenient for the particular forming process.

Most metals, with the exception of some precious ones, are found in the form of minerals or ores. The ores are smelted to convert them into metals; for example, iron is obtained from iron ore (haematite), aluminium from bauxite, and copper from copper pyrites.

Iron ore together with other elements is smelted in a blast furnace to give pig iron. According to the type of plant, the molten pig iron is cast as pigs or is transferred to a steel-making process. Cast pigs are refined in a cupola to give cast iron which is cast as notched ingots or bars of relatively small cross-section for ease of remelting in the foundry as required. Pig iron taken in a molten state to the steel-making process is made into steel which is cast as slabs for subsequent rolling, drawing, or forging.

Aluminium is extracted from bauxite by an electrolytic process. Commercially pure aluminium is soft and weak and it is alloyed to improve the mechanical properties. Aluminium alloys are available wrought or cast, in sections convenient for subsequent working by rolling, drawing, casting, or extrusion.

Copper is extracted from copper pyrites and is refined by remelting, in a furnace or electrolytically. Copper is alloyed to produce a range of brasses and bronzes. These materials may then be rolled, drawn, cast, or extruded.

2.1 Forms of supply of raw materials

Specific objective To describe the form of supply of the raw materials for sand casting, rolling, extrusion, drawing, and forging.

Where the metal has to be remelted, as in the case of casting, it is usual to supply it to the foundry in the form of notched ingots or bars of relatively small cross-section. These can be broken into smaller pieces for ease of handling and loading into small furnaces.

Where the metal is to be subjected to further forming processes – e.g. rolling, extrusion, drawing, or forging – it would be wasteful to remelt, both from an energy viewpoint and from the effect the remelting would have on the physical and mechanical properties. In this case the raw material would be supplied in the form most convenient for the process – i.e. slabs (width greater than three times the thickness) for rolling into sheet; blooms and billets (smaller section) for rolling into bar, sections, etc. and for forging and extrusion. Hot-rolled rod is supplied for cold drawing into rod, tube, and wire.

2.2 Properties of raw materials

Specific objective To state the properties required of a raw material used in sand casting, rolling, extrusion, drawing, or forging.

Fluidity This property is a requirement for a metal which is to be cast. The metal must flow freely in a molten state in order to completely fill the mould cavity.

Ductility A ductile material can be reduced in cross-section without breaking. Ductility is an essential property when drawing, since the material must be capable of flowing through the reduced diameter of the die and at the same time withstand the pulling force. The reduction in cross-sectional area aimed for in a single pass through the die is usually between 25% and 45%.

Malleability A malleable material can be rolled or hammered permanently into a different shape without fracturing. This property is required when rolling and forging.

Plasticity This is a similar property to malleability, involving permanent deformation without fracture. This property is required in forging and extrusion, where the metal is rendered plastic, i.e. made more pliable, by the application of heat.

Toughness A material is tough if it is capable of absorbing a great deal of energy before it fractures. This property is required when forging.

2.3 Sand casting

Specific objective   To describe with the aid of diagrammatic sketches the basic steps of sand casting.

Casting is the simplest and most direct way of producing a finished shape from metal. Casting shapes from liquid metals can be done by a variety of processes, the simplest of which is sand casting.

For the production of small castings, a method known as box moulding is employed. The box is made up of two frames with lugs at each end into which pins are fitted to ensure accurate alignment when the frames are placed together. The top frame is referred to as the cope and the bottom frame as the drag.

A removable pattern is used to create the required shape of cavity within the mould. The pattern is made in two parts, split at a convenient position for ease of removal from the mould – this position being known as the parting line. The two parts of the pattern are accurately aligned with each other using pins or dowels.

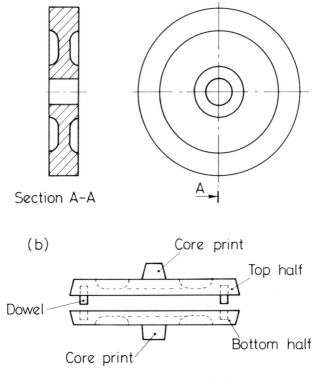

Fig. 2.1 Gear blank and pattern

Consider the gear blank shown in fig. 2.1(a), the pattern for which is shown in fig. 2.1(b). One half of the pattern is placed on a board. The drag is then placed on the board such that the pattern half is roughly central. Moulding sand is then poured into the drag and is pressed firmly against the pattern. The drag is then filled completely and the sand is firmly packed using a rammer, fig. 2.2(a). The amount of ramming should be sufficient for the sand to hold together but not enough to prevent the escape of gases produced during pouring of the molten metal. After ramming is complete, the sand is levelled off flush with the edges of the drag. Small vent holes can be made through the sand to within a few millimetres of the pattern, to assist the escape of gases.

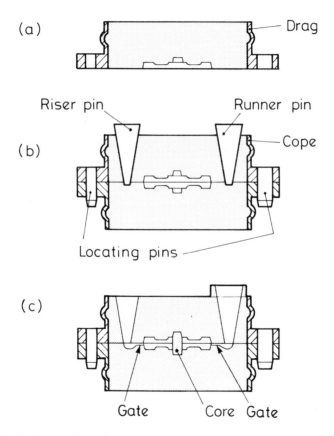

Fig. 2.2 Basic steps in sand casting

The drag is then turned over and the second half pattern is located by means of the dowels. The upturned surface is then covered with a fine coating of dry 'parting sand', to prevent bonding between the sand in the cope and that in the drag. The cope is accurately positioned on the drag by means of the locating pins. To allow entry for the molten metal, a tapered plug known as a runner pin is placed to one side of the pattern. A second tapered plug known as a riser pin is placed at the opposite side of the pattern, fig. 2.2(b) – this produces an opening which, when filled with molten metal during pouring, provides a supply of hot metal to compensate for shrinkage as the casting cools. The cope is then filled, rammed, and vented as for the drag. The pins are then removed and the top of the runner hole is enlarged to give a wide opening for pouring the metal.

The cope is then carefully lifted off and turned over. Both halves of the pattern are carefully removed. Small channels known as gates are then cut from the bottom of the runner and riser to enable metal to fill the mould cavity. Any loose sand is blown away to leave a clean cavity in the mould. The core is placed in position in the bottom half of the mould box (the drag) and the top mould box (the cope) is carefully replaced in position with the aid of the locating pins and is clamped to prevent lifting when the metal is poured, fig. 2.2(c). The core which is necessary to provide a hollow section, in this case the bore of the gear blank, is made separately.

The mould is now ready for pouring. When the metal has solidified, the mould is broken up to release the casting. The runner and riser are broken off, and the rough edges are removed by fettling (i.e. hand grinding).

Patterns

For the production of small quantities of castings, patterns are made from wood, smoothed, painted, or varnished to give a smooth finish to the casting. Patterns are made larger than the finished part, to allow for shrinkage of the casting when it cools. A special rule, known as a contraction rule, is available to suit different metals – the pattern-maker makes the pattern using measurements from the contraction rule, which automatically gives the correct dimension of pattern with due allowance for shrinkage whatever the size of dimension.

Some surfaces of a casting may require subsequent machining, e.g. surfaces requiring a greater accuracy of size, flatness, or surface finish than can be achieved by casting. Extra metal must be left on these surfaces, and the amount to be left for removal by machining must be allowed for on the appropriate surface of the pattern.

Where cores are to be incorporated in a casting (see below), provision must be made on the pattern to provide a location seating in the mould. These sections added to the pattern are known as core prints.

To allow the pattern to be easily removed from the mould, a small angle or taper known as draft is incorporated on all surfaces perpendicular to the parting line.

Sand

Moulding sand must be permeable, i.e. porous, to allow the escape of gases and steam; strong enough to withstand the mass of molten metal; resist high temperatures; and have a grain size suited to the desired surface of the casting.

Silica sand is used in moulding, the grains of sand being held together in different ways.

In green-sand moulds the grains are held together by moist clay, and the moisture level has to be carefully controlled in order to produce satisfactory results.

Dry-sand moulds start off in the same way as green-sand moulds but the moisture is driven off by heating after the mould has been made. This makes the mould stronger and is suited to heavier castings.

With CO_2 (carbon dioxide) sand the silica grains are coated with sodium silicate instead of clay. When the mould is made, it is hardened by passing carbon-dioxide gas through it for a short period of time. The sand 'sets' but is easily broken after casting.

Specially prepared facing sand is used next to the pattern to give an improved surface to the casting. The mould can then be filled using a backing sand.

Cores

When a casting is to have a hollow section, a core must be incorporated into the mould. Cores are normally of green sand or dry sand, made separately in a core box, and inserted in the mould after the pattern is removed and before the mould is closed. They are located and supported in the mould in a seating formed by the core prints on the pattern. The core must be strong enough to support itself and withstand the flow of molten metal, and in some cases it may be necessary to reinforce it with wires to give added strength.

The more complex cores are produced from CO_2 sand.

2.4 Rolling

Specific objective To describe with the aid of diagrammatic sketches the basic steps of rolling.

The cast ingots produced by the raw-material producers are of little use for manufacturing processes until they have been formed to a suitable shape, i.e. sheet, plate, strip, bar, sections, etc.

One of the ways in which shapes can be produced in order that manufacturing processes can subsequently be carried out is by rolling. This can be performed as hot rolling or cold rolling – in each case, the metal is worked while in a solid state and is shaped by plastic deformation.

The reasons for working metals in their solid state are firstly to produce shapes which would be difficult or expensive to produce by other methods, e.g. long lengths of sheet, section, rods, etc., and secondly to improve mechanical properties.

The initial stage of converting the ingot to the required shape is by hot rolling. During hot working, the metal is in a plastic state and is readily formed by pressure as it passes through the rolls. Hot rolling has a number of other advantages:

a) Most ingots when cast contain many small holes – a condition known as porosity. During hot rolling, these holes are pressed together and eliminated.
b) Any impurities contained in the ingot are broken up and dispersed throughout the metal.
c) The internal grain structure of the metal is refined, resulting in an improvement of the mechanical properties, e.g. ductility and strength.

Hot rolling does, however, have a number of disadvantages. Due to the high temperatures, the surface oxidises – producing a scale which results in a poor surface finish, making it difficult to maintain dimensional accuracy. Where close dimensional accuracy and good surface finish are not of great importance, e.g. structural shapes for construction work, a descaling operation is carried out and the product is used as-rolled. Alternatively, further work can be carried out by cold rolling.

When metal is cold rolled, greater forces are required, necessitating a large number of stages before reaching the required shape. The strength of the material is greatly improved, but this is accompanied by a decrease in ductility. Depending on the number of stages required in producing the shape, annealing (or softening) may have to be carried out between stages. Besides improving mechanical properties, cold rolling produces a good surface finish with high dimensional accuracy.

In the initial stage of converting it to a more suitable form, the ingot is first rolled into intermediate shapes such as blooms, billets, or slabs. A bloom has a square cross-section with a minimum size of 150 mm square. A billet is smaller than a bloom and may have a square cross-section from 40 mm up to the size of a bloom. A slab is rectangular in cross-section with a minimum width of 250 mm and a minimum thickness of 40 mm. These are then cut into convenient lengths for further hot or cold working.

Most primary hot rolling is carried out in either a two-high reversing mill or a three-high continuous mill.

In the two-high reversing mill, fig. 2.3, the metal is passed between the rolls in one direction. The rolls are then stopped, closed together by an amount depending on the rate of reduction required, and reversed, taking the material back in the opposite direction. This is repeated, with the rolls

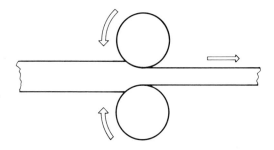

Fig. 2.3 Two-high reversing mill

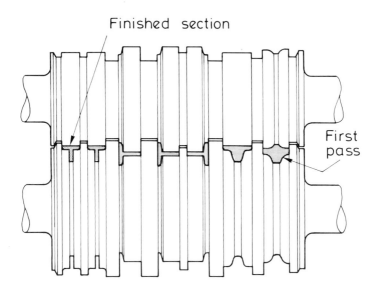

Fig. 2.4 Rolls for producing a tee section

closed a little more each time, until the final size of section is reached. At intervals throughout this process, the metal is turned on its side to give a uniform structure throughout. Grooves are provided in the top and bottom rolls to give the various reductions and the final shape where appropriate, fig. 2.4.

In the three-high continuous mill, fig. 2.5, the rolls are constantly rotating, the metal being fed between the centre and upper rolls in one direction and between the centre and lower rolls in the other. A platform is posi-

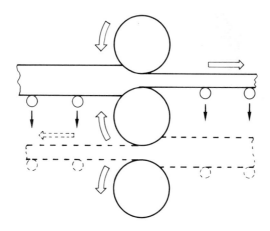

Fig. 2.5 Three-high continuous mill

tioned such that it can be raised to feed the metal through in one direction or support it coming out from between the rolls in the opposite direction or be lowered to feed the metal back through or to support it coming out from between the rolls in the opposite directions.

In cold rolling, the roll pressures are much greater than in hot rolling – due to the greater resistance of cold metal to reduction. In this case it is usual to use a four-high mill, fig. 2.6. In this arrangement, two outer rolls of large diameter are used as back-up rolls to support the smaller working rolls and prevent deflection.

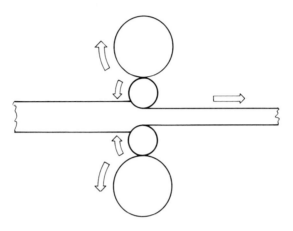

Fig. 2.6 Four-high mill

When rolling strip, a series of rolls are arranged in line and the strip is produced continuously, being reduced by each set of rolls as it passes through before being wound on to a coil at the end when it reaches its final thickness.

2.5 Extrusion

Specific objective   To describe with the aid of diagrammatic sketches the basic steps of extrusion.

Extrusion usually has to be a hot-working process, due to the very large reduction which takes place during the forming process. In operation, a circular billet of metal is heated to render it plastic and is placed inside a container. Force is then applied to the end of the billet by a ram which is usually hydraulically operated. This applied force pushes the metal through an opening in a die to emerge as a long bar of the required shape. The extrusion produced has a constant cross-section along its entire length. The die may contain a number of openings, simultaneously producing a number of extrusions.

There are a number of variations of the extrusion process, two common methods being direct extrusion and indirect extrusion.

Direct extrusion

This process, fig. 2.7, is used for the majority of extruded products. A heated billet is positioned in the container with a dummy block placed behind it. The container is moved forward against a stationary die and the ram pushes the metal through the die. After extrusion, the container is moved back and a shear descends to cut off the butt end of the billet at the die face. The process is repeated with a new billet. As the outside of the billet moves along the container liner during extrusion, high frictional forces have to be overcome, requiring the use of high ram forces.

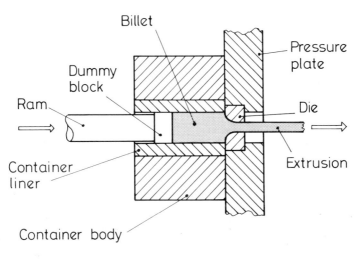

Fig. 2.7 Direct extrusion

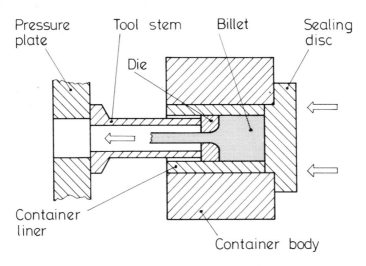

Fig. 2.8 Indirect extrusion

Indirect extrusion
In this process, fig. 2.8, the heated billet is loaded in the container which is closed at one end by a sealing disc. The container is moved forward against a stationary die located at the end of a hollow stem. The extrusion passes from the die through the inside of the hollow stem. Because the container and billet move together and there is no relative movement between them, friction is eliminated. As a result, longer and larger-diameter billets can be used than with direct-extrusion presses of the same power. Alternatively lower forces are required with the same size billets.

The indirect-extrusion process does have limitations, however. Metal flow tends to carry surface impurities into the extruded metal, and the billets have to be machined or chemically cleaned. The size of the extrusion is limited to the inside diameter of the hollow stem, and die changing is more cumbersome than with direct extrusion.

The important features of the extrusion process are

a) the complexity of shape possible is practically unlimited, and finished products can be produced directly, fig. 2.9;

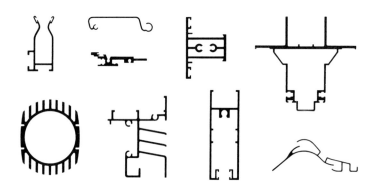

Fig. 2.9 Examples of extruded shapes

b) a good surface finish can be maintained;
c) good dimensional accuracy can be obtained;
d) large reductions in cross-sectional area can be achieved;
e) the metal is in compression during the process, so relatively brittle materials can be extruded;
f) the mechanical properties of the material are improved.

The extrusion process is, however, limited to products which have a constant cross-section. Any holes, slots, etc. not parallel to the longitudinal axis have to be machined. Due to extrusion-press power capabilities, the size of shape which can be produced is limited. The process is normally limited to long runs, due to die costs, but short runs can be economical with simple die shapes. Typical materials used are copper and aluminium and their alloys.

2.6 Drawing

Specific objective To describe with the aid of diagrammatic sketches the basic steps of drawing.

The primary process of drawing is a cold-working process, i.e. carried out at room temperature. It is mainly used in the production of wire, rod, and bar.

Wire is made by cold drawing a previously hot-rolled rod through one or more dies, fig. 2.10, to decrease its size and improve the physical properties. The hot-rolled rod – usually around 10 mm in diameter – is first cleaned in an acid bath to remove scale, a process known as pickling. This ensures a good finish on the final drawn wire. The rod is then washed with water to remove and neutralise the acid. The end of the rod is then pointed so that it can be passed through the hole in the die and be gripped in a vice attached to the drawing machine. The rod is then pulled through the die to give the necessary reduction in section.

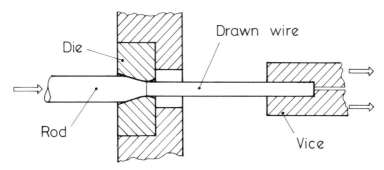

Fig. 2.10 Wire drawing

In continuous wire drawing, the wire is pulled through a series of progressively smaller dies until the final-size section is reached.

The strength of the material will limit the force which can be applied in pulling the wire through the die, while the ductility of the material limits the amount of reduction possible through each die. Typical materials used are steel, copper, aluminium, and their alloys.

2.7 Forging

Specific objective To describe with the aid of diagrammatic sketches the basic steps of forging.

Forging is a hot-working process, heat being necessary to render the metal plastic in order that it may be more easily shaped. The oldest form of forging is hand forging as carried out by the blacksmith. Hand tools are used to manipulate the hot metal to give changes in section and changes in shape by bending, twisting, etc. Due to the hand operation, it is not possible to achieve high degrees of accuracy or extreme complexity of shape. This method is limited to one-off or small quantities and requires a high degree of skill.

When forgings are large, some form of power is employed. Steam or compressed-air hammers or a forging press is used, the process being known as open-sided forging. In this process, the hot metal is manipulated to the required shape by squeezing it between a vertically moving die and a stationary die attached to the anvil, fig. 2.11. This method of forging is carried out under the direction of a forge-master who directs the various stages of turning and moving along the length until the finished shape and size are obtained. Again, great skill is required. This method is used to produce large forgings such as propeller shafts for ships.

When large quantities of accurately shaped products are required, these are produced by a process known as closed-die forging or drop forging. With this method, the hot metal is placed between two halves of a die each containing a cavity such that when the metal is squeezed into the cavity a completed forging of the required shape is produced, fig. 2.12. The metal is subjected to repeated blows, usually from a mechanical press, to ensure proper flow of the metal to fill the die cavity. A number of stages through a series of dies may be required, each stage changing the shape gradually until the final shape is obtained. The number of stages required depends on the size and shape of the part, how the metal flows, and the degree of accuracy required.

One of the two halves of the die is attached to the moving part of the press, the other to the anvil. The cavity contained in the die is designed such that the parting line enables the finished forging to be removed and incorporates a draft in the direction of die movement in the same way as do

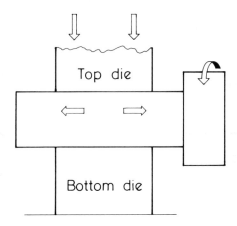

Fig. 2.11 Open-die forging

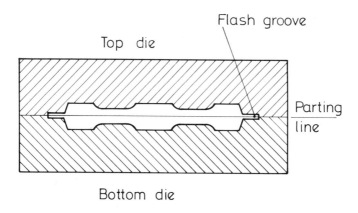

Fig. 2.12 Closed-die forging

the pattern and mould in sand casting. The size of cavity also allows for additional material on faces requiring subsequent machining.

Since it is impossible to judge the exact volume required to just fill the die cavity, extra metal is allowed for and this is squeezed out between the two die halves as they close. This results in a thin projection of excess metal on the forging at the parting line, known as flash. This flash is removed after forging by a trimming operation in which the forging is pushed through a correct-shape opening in a die mounted in a press.

Forging is used in the production of parts which have to withstand heavy or unpredictable loads, such as levers, cams, gears, connecting rods, and axle shafts. Mechanical properties are improved by forging, as a result of the flow of metal being controlled so that the direction of grain flow increases strength. Figure 2.13 shows the difference in the grain flow between a shaft with a flange which has been forged up, fig. 2.13(a), and one machined from a solid rolled bar, fig. 2.13(b). Any form machined on the flange, such as gear teeth, would be much weaker when machined from solid bar. The structure of the material is refined due to the hot working, and the density is increased due to compression forces during the forging process.

The process of drop forging is normally restricted to larger batch quantities, due to die costs.

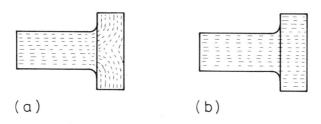

Fig. 2.13 Direction of grain flow

2.8 Selection of a primary process

Specific objective To select a suitable primary process which might be employed to produce a given component.

A number of factors have to be considered before a choice of process for a given component can be made. For example, consideration would have to be given to the type of material to be used, the mechanical properties required, shape, accuracy, degree of surface finish, and the quantity to be produced. Some of these factors are shown in Table 2.1 together with the appropriate primary processes, as an aid to the correct choice. For example, the requirement of variation of shape in three-dimensions would eliminate rolling, extrusion, and drawing; a high degree of accuracy would eliminate sand casting and forging; and so on.

Table 2.1 Features of primary forming processes

	Sand casting	Rolling		Extrusion	Drawing	Forging
		Hot	Cold			
Improved mechanical properties		✓	✓	✓	✓	✓
Three-dimensional shape variation	✓					✓
Constant cross-section		✓	✓	✓	✓	
Large quantities		✓	✓	✓	✓	✓
Good surface finish			✓	✓	✓	
High dimensional accuracy			✓	✓	✓	
High tool costs				✓		✓

2.9 Hazards in primary process work

Specific objective To identify the main hazards in primary process work and to outline the necessary precautions.

It is in your interests to be aware of the potential hazards to which you may be exposed and the possible effects to your health and safety. Any instruction on health and safety must be treated seriously. Safety signs are displayed for your benefit and must be complied with at all times.

BS 5378: part 1:1980, 'Safety signs and colours. Specification for colour and design', for example, was introduced to achieve standardisation of communication and awareness of safety colours and shapes. Through the use of shape, colour, and symbols, the aim is to draw attention to the objects which affect health and safety. The signs were covered in *Workshop processes, practices, and materials* and it is sufficient here to 'remind' you that the colour red is used to indicate danger, yellow to indicate caution, and green for go or a safe condition, as in a traffic-light system. An additional colour – blue – is used to indicate where some form of protection is required, e.g. the use of helmets, ear protection, etc.

Where there are known risks to health and safety, steps should be taken to control them. Methods of control would include

a) substitution of a less hazardous process;
b) enclosure of the process;
c) isolation of the process from other people, with personal protection for exposed persons;
d) dust suppression or extraction, e.g. by a ventilation system;
e) personal protective clothing and equipment;
f) short working periods, to lessen daily exposure;
g) maintenance of general environmental tidiness and cleanliness;
h) health education and warning notices.

The following hazards and precautions are outlined with primary-process work in mind. However, many are also applicable to general industry.

Personal protection

In hot-working processes involving molten metal – especially in casting – it is necessary to use some form of personal protection. Gauntlets, gaiters, aprons, and footwear for protection against burns incurred from molten-metal splashes and from hot sand and from impact risks must be worn. The footwear is specially designed and incorporates a quick-release device enabling it to be removed rapidly. Eye protection from molten metal is essential, and approved faceshields for molten metal must be worn. Safety footwear incorporating impact-resistant toe caps must always be worn where there is a risk of falling objects, e.g. billets, castings, forgings, and equipment. Different materials are used for the sole to give resistance against oil or heat and to prevent slipping. Safety helmets are essential where there is any risk from overhead. Remember, protective clothing is provided for your benefit and must be worn at all times.

Noise

A great many workers are exposed to hazardous noise levels at their workplace. Hearing is affected not only by high noise levels but also by continual exposure to much lower levels. The plant and machinery used in primary processing such as steel production, rolling, forging, and in the foundry can produce hazardous noise.

The best method of avoiding this is to reduce the noise at its source by the use of screens and enclosures. When this is not possible, however, hearing protectors must be provided and worn. The prime function of hearing protectors is to reduce the noise level at the wearer's ears to within safe limits. The two main classes of protector are ear plugs inserted inside the ear and ear muffs fitting around the ears and pressed against the sides of the head. To avoid the risk of partial or total deafness, it is essential to wear hearing protectors in any area where hazardous noise levels exist.

Dust and fumes
Where a process produces dust or fumes, these may be kept at acceptable levels by the use of a ventilation system. Alternatively, if a hazard to the operator is present, some form of mask will be necessary which must be provided and worn.

Fire
Where hot or molten metals are being worked, fire is a potential hazard. Flammable materials should not therefore be used or stored in close proximity to these processes. Simple items such as waste rags or sacks impregnated with oil have been known to cause fires through spontaneous ignition. Dust-laden atmospheres can ignite and result in explosion and fire. Ensure that you know where the fire-fighting appliances are located, which one to use for a particular type of fire, and how it should be used. Get to know the fire drill and the means of escape from the building.

Environment
Many accidents result from falls brought about by obstructions on the factory floor, lack of cleanliness of a floor, or a bad floor surface and as a result of bad lighting. Always keep the floor clean by wiping up spilt oil, loose sand, etc. and never obstruct the floor with equipment or parts.

Moving machinery
Any part of moving machinery is a potential hazard and should be guarded. In most primary processes, large machinery is used – e.g. rolling mills, forging and extrusion presses, etc., as well as conveyors – and great care and attention to the appropriate warnings and safety procedures is essential. All guards must be in position at all times.

Materials handling
Manual handling is a major cause of injury. Done incorrectly, it usually results in some form of back injury. The correct procedure, as discussed in *Workshop processes, practices, and materials*, should always be followed. Wherever possible, use handling aids such as wheeled crowbars, sack trucks, or skates.

In primary processing, the materials being handled are often hot and of large mass and require the use of mechanical equipment such as cranes, lifting tackle, and industrial trucks. These provide a host of potential hazards. Cranes and their hooks, chains, and lifting tackle are subject to strict safety regulations requiring registers of certificates of test and examination and records of examination. It is prohibited to operate a crane in excess of its safe-working-load (SWL) capacity, which must be notified to operators and displayed. Hooks should either be provided with an efficient device to prevent displacement of the load off the hook or be of such a shape as to avoid the risk of displacement. A 'C' hook or one fitted with a safety catch is recommended. All persons working with cranes must be trained in the use of accepted hand signals.

Lifting appliances such as ropes, chains, and slings must be properly made and strong enough for the work. They must not be used until they have been tested and examined, and a certificate of the test and examination, showing the safe working load, must be obtained for each item. They must never be overloaded.

Industrial trucks such as fork lifts are commonly used. No person should be permitted to operate a fork-lift truck unless specifically selected, trained, and authorised to do so. The vehicle's condition must be checked each day, and the vehicle must be operated in adequate working conditions, i.e. clearly marked aisles and suitable surfaces with appropriate traffic controls. Trucks should be operated at a safe speed and give warning when approaching doorways, blind corners, etc. by the use of horns. Always ensure that loads are secure, and check the condition of pallets before use. Safe procedures for stacking of loads should always be followed.

Emergency procedure

If you are at the site of an accident, it is essential that you keep calm. Immediately call the emergency services. Do not move the victim unless he or she is in immediate danger. Put burning victims on the ground wrapped in a coat. Keep the victim warm and as comfortable as possible until the emergency services arrive. In your own interests as well as those of your colleagues, learn first aid by attending an approved course and learn how to deal with electric shock – it could save a life.

Exercises on chapter 2

1 State an advantage and a disadvantage of hot rolling.
2 Name two types of extrusion process.
3 The top frame of the box used in sand casting is called the_____.
4 What is the purpose of a core in sand casting?
5 Define toughness, malleability, and ductility.
6 What is the purpose of the two outer rolls used in a four-high mill?
7 State the reason for using parting sand between the two halves of the moulding box in sand casting?
8 Briefly describe the basic steps of the drawing process.
9 State four important features of the extrusion process.
10 State the aim of the use of safety colours and safety signs.
11 What is 'draft' and why is it used on the pattern in sand casting?
12 State four steps which can be taken to control known risks to health and safety in primary-process work.
13 Drop forgings have a thin projection of excess metal around the outer edge. What is it called and how is it removed?
14 List the steps you would take if you were present at the site of an accident.

3 Presswork

General objective The student understands the principles of simple presswork.

The term 'presswork' is used here to describe the process in which force is applied to sheet metal with the result that the metal is cut, i.e. in blanking and piercing, or is formed to a different shape, i.e. in bending.

The pressworking process is carried out by placing the sheet metal between a punch and die mounted in a press. The punch is attached to the moving part – the slide or ram – which applies the necessary force at each stroke. The die, correctly aligned with the punch, is attached to the fixed part or bedplate of the press.

The press used may be manually operated by hand or by foot and used for light work or it may be power operated, usually by mechanical or hydraulic means, and capable of high rates of production.

The time to produce one component is the time necessary for one stroke of the press slide plus load/unload time or time for feeding the material. Using a power press, this total time may be less than one second.

It is possible to carry out a wide range of operations in a press, and these include blanking, piercing, and bending.

Blanking is the production of an external shape, e.g. the outside diameter of a washer.

Piercing is the production of an internal shape, e.g. the hole in a washer.

Bending – in this case simple bending – is confined to a straight bend across the metal sheet in one plane only.

3.1 Presses

Specific objective To describe the purpose, advantages, and limitations of the fly press and the power press.

The fly press

Blanking, piercing, and bending of light work where the required force is small and the production rate is low may be carried out on a fly press.

A hand-operated bench-type fly press is shown in fig. 3.1. The body is a C-shaped casting of rigid proportions designed to resist the forces acting during the pressworking operation. The C shape gives an adequate throat depth to accommodate a range of work sizes. (The throat depth is the distance from the centre of the slide to the inside face of the body casting.) The bottom part of the casting forms the bed to which tooling is attached.

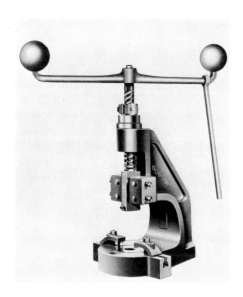

Fig. 3.1 Fly press

The top part of the casting is threaded to accept the multi-start square-threaded screw which carries the handle at its top end and the slide at its bottom end. An adjustable threaded collar is fitted at the top end of the screw and can be locked at a required position to avoid overtravel of the screw during operation. The slide contains a hole and a clamping screw to locate and secure the punch. The horizontal portion of the handle is fitted with ball masses which produce a flywheel effect when larger forces are required. These masses are fitted on spikes, and one or both may be removed when smaller forces are required.

In operation, the vertical handle is grasped and the handle is partially rotated. This provides, through the multi-start thread, a vertical movement of the slide. A punch and die fitted in the slide and on the bed, and correctly aligned with each other, are used to carry out the required press-working operation.

The fly press can be set up easily and quickly for a range of blanking, piercing, and bending operations, and the manual operation gives a greater degree of sensitivity than is often possible with power presses. This type of press may also be used for operations such as pressing dowels and drill bushes into various items of tooling. Due to the manual operation, its production rates are low.

Power presses
Power presses are used where high rates of production are required. A power press may be identified by the design of the frame and its capacity – i.e. the maximum force capable of being delivered at the work, e.g.

500 kN ('50 tons'). The source of power may be mechanical or hydraulic. Different types of press are available in a wide range of capacities, the choice depending on the type of operation, the force required for the operation, and the size and type of tooling used.

One of the main types of power press is the open-fronted or gap-frame type. This may be rigid or inclinable. The inclinable feature permits finished work to drop out the back by gravity. A mechanical open-fronted rigid press is shown in fig. 3.2. The model shown has a capacity of 1000 kN ('100 tons') and operates at 60 strokes per minute (it is shown without guards, for clarity). The open nature of this design gives good accessibility of the tools and allows the press to be operated from either side or from the front.

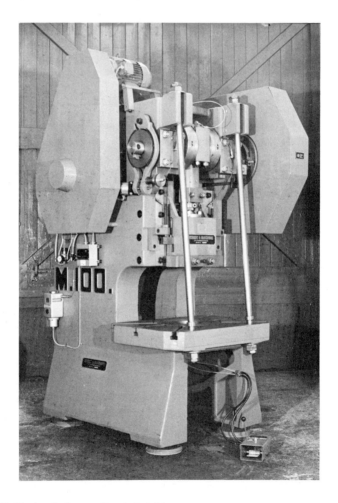

Fig. 3.2 Mechanical open-fronted rigid power press

The limitation of the open-fronted press is the force which can be applied. High forces have a tendency to flex the frame and so open the gap between the slide and the bedplate. This flexing of the frame can be overcome on the large-capacity presses by fitting tie-rods between the bedplate and the top of the frame as shown in fig. 3.2.

As the capacity of a press is increased, it becomes necessary to increase the strength and rigidity of its frame – for the reason already outlined. This is achieved in the other main type of power press – the straight-sided or column type – since the large forces are taken up in a vertical direction by the side frames. A mechanical straight-sided press of 3000 kN ('300 tons') capacity is shown in fig. 3.3.

Fig. 3.3 Mechanical straight-sided power press

Straight-sided presses sacrifice adaptability and accessibility to gain frame rigidity and are best suited to work on heavy-gauge metals and on large surfaces. Since the sides are closed by the side frames and are open at the front and the back, these presses are limited to operation from the front only.

Mechanical presses Mechanical power presses derive their energy for operation from a constantly rotating flywheel driven by an electric motor. The flywheel is connected to a crankshaft through a clutch which can be set for continuous or single stroking. In the single-stroke mode, the clutch is automatically disengaged at the end of each stroke and the press will not restart until activated by the operator, e.g. by operating a foot pedal. A connecting rod is attached at one end to the crankshaft and at its other end to the slide. Adjustment is provided to alter the position of the slide, and in some presses the length of stroke can also be adjusted by means of an eccentric on the crankshaft. A brake is fitted to bring the crankshaft to rest at the correct position.

With mechanical presses, the maximum force is available at the bottom of the slide stroke. In blanking and piercing operations, the work is done very near to the bottom of the stroke. However, where a part is blanked and then taken further down the stroke, e.g. to form a bend, less force will be available for the blanking operation which will have been carried out some distance above the bottom of the stroke. For example, a 500 kN press with a stroke of 120 mm will exert a force of only 120 kN half-way down its stroke.

Hydraulic presses Hydraulic power presses derive their power from high-pressure hydraulic pumps which operate the ram. The load applied is completely independent of the length and stroke, i.e. full load can be applied at any point in the stroke. The applied load is controlled by a relief valve which gives automatic protection against overload. Again by means of valves, the ram can be made to approach the work rapidly and then be shifted to a lower speed before contacting the work, thus prolonging the life of the tool but still giving fast operating speeds. Rapid ram reversal can also be controlled. Switches are incorporated to determine the positions at which these controls become effective, thereby increasing productivity by making tool setting faster and by keeping the actual working stroke to a minimum.

The number of moving parts are few and these are fully lubricated in a flow of pressurised oil, leading to lower maintenance costs. Fewer moving parts and the absence of a flywheel reduce the overall noise level of hydraulic presses compared to mechanical presses.

Longer strokes are available than with mechanical presses, giving greater flexibility of tooling heights.

A typical hydraulic press of 100 kN ('10 ton') capacity is shown in fig. 3.4. This model has a ram advance speed of 475 mm/s, a pressing speed of 34 mm/s, and a return speed of 280 mm/s. The model shown is

Fig. 3.4 Hydraulic straight-sided power press

fitted with a light-screen guard operated by a continuous curtain of infra-red light.

Safety
Many serious accidents to operators and tool setters have occurred in the use of power presses. Stringent safety requirements must be met in the use of power presses, and these are covered by the Power Presses Regulations. The regulations include requirements for the thorough examination and testing by a competent person of power-press mechanisms and safety devices after installation, before use, and periodically as a condition of use. A vital factor in the prevention of accidents at power presses is the effective maintenance of both presses and guards in sound working condition.

Power-press mechanisms

Press mechanisms such as the clutch, brake, connecting rod, and flywheel journals, as well as guards and guard mechanisms, must be subjected to systematic and regular thorough examination. Many presswork operations involve feeding and removing workpieces by hand, and every effort must be made to ensure that this can be done safely without the risk that the press will operate inadvertently while the operator's hands are within the tool space. The most obvious way of avoiding accidents to the operator is to design the tools in such a way as to eliminate the need for the operator to place his fingers or hands within the tool space for feeding or removal of work. This can often be done by providing feeding arrangements such that the operator's hands are outside the working area of the tools.

All dangerous parts of the press must be guarded, and the four principal ways of guarding the working area are

i) enclosed tools, where the tools are designed in such a way that there is insufficient space for entry of fingers;
ii) fixed guards, which prevent fingers or hands reaching into the tool space at any time;
iii) interlocked guards, which allow access to the tools but prevent the clutch being engaged until the guards are fully closed – the guards cannot be opened until the cycle is complete, the clutch is disengaged, and the crankshaft has stopped;
iv) automatic guards, which push or pull the hand clear before trapping can occur;
v) light-screen guards – operated by a continuous curtain of infra-red light which, if broken, stops the machine.

3.2 Press-tool design

Specific objective   To explain the design and use of punches, dies, and bolsters.

The tools used in presses are punches and dies. The punch is attached to the press slide and is moved into the die, which is fixed to the press bed-plate. In blanking and piercing, the punch and die are the shape of the required blank and hole and the metal is sheared by passing the punch through the die. In bending, the punch and die are shaped to the required form of the bend and no cutting takes place. In each case, the punch and die must be in perfect alignment. The complete assembly of punch and die is known as a press tool.

Blanking and piercing

In blanking and piercing operations the work material, placed in the press tool, is cut by a shearing action between the adjacent sharp edges of the punch and the die. As the punch descends on to the material, there is an initial deformation of the surface of the material followed by the start of

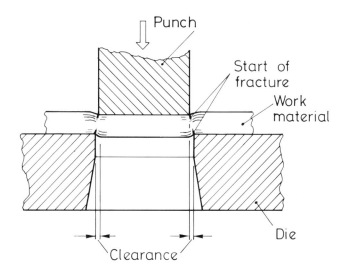

Fig. 3.5 Shearing action of punch and die

fracture on both sides, fig. 3.5. As the tensile strength of the material is reached, fracture progresses and complete failure occurs.

The shape of the side wall produced by this operation is not straight as in machining operations and is shown greatly exaggerated in fig. 3.6. The exact shape depends on the amount of clearance between the punch and the die. Too large a clearance leads to a large angle of fracture and a large burr, while too small a clearance will result in premature wear of the tools and a risk of tool breakage.

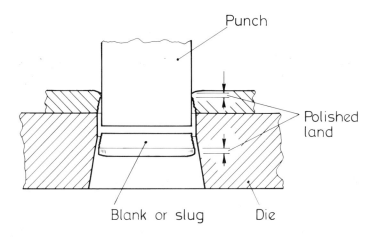

Fig. 3.6 Characteristics of sheared edge

The clearance or the space between the punch and the die is quoted as a percentage of the work-material thickness per side. Establishment of the correct clearance to be used for a given blanking or piercing operation is influenced by the required characteristics of the cut edge and by the thickness and properties of the work material. The values given in Table 3.1 are offered only as a general guide.

Table 3.1 Typical values of clearance for press-tool design

Work material	Clearance per side (% of work-material thickness)
Low-carbon steel	5–7
Aluminium alloys	2–6
Brass:	
annealed	2–3
half hard	3–5
Phosphor bronze	3.5 5
Copper:	
annealed	2–4
half hard	3–5

In blanking and piercing operations, the punch establishes the size of the hole and the die establishes the size of the blank. Therefore in piercing, where an accurate size of hole is required, the punch is made to the

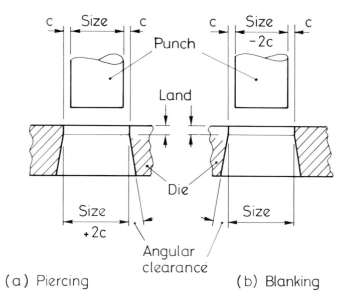

Fig. 3.7 Clearance on punch and die for piercing and blanking operations

required hole size and the clearance is made on the die, fig. 3.7(a). Conversely, in blanking, where an accurate size of blank is required, the die is made to the required blank size and the clearance is made on the punch, fig. 3.7(b). On this basis, the material punched through during piercing is scrap and the material left behind in the die during blanking is scrap.

Example 1 It is required to produce 50 mm diameter blanks from 2 mm thick low-carbon steel. If a clearance of 6% is chosen, then

diameter of die = diameter of blank = 50 mm

clearance per side = 6% of 2 mm = 0.12 mm

clearance on diameter is therefore 2 × 0.12 mm = 0.24 mm

Thus the diameter of the punch is smaller than the die by this amount; therefore

diameter of punch = 50 mm − 0.24 mm = 49.76 mm

Example 2 It is required to punch 20 mm diameter holes in 1.5 mm thick copper. If a clearance of 4% is chosen, then

diameter of punch = diameter of hole = 20 mm

clearance per side = 4% of 1.5 mm = 0.06 mm

clearance on diameter is therefore 2 × 0.06 mm = 0.12 mm

Thus the diameter of the die is larger than the punch by this amount; therefore

diameter of die = 20 mm + 0.12 mm = 20.12 mm

To prevent the blanks or slugs removed by the punch from jamming in the die, it is usual to provide an angular clearance below the cutting edge of the die as shown in fig. 3.7(a). Thus the pieces can fall through the die, through a hole in the bedplate, into a bin. A land equal in width to approximately twice the metal thickness may be provided, which enables a large number of regrinds to be carried out on the top die face to maintain a sharp cutting edge.

Stripping
As the punch enters the work material during a blanking or piercing operation, it becomes a tight fit in the material. To withdraw the punch without lifting the material along with the punch, it is necessary to provide a method of holding the material while the punch is withdrawn on the upward stroke. This is known as stripping and can be done by providing a fixed stripper or a spring-type stripper.

A fixed stripper, fig. 3.8, is used when the work material is in the form of strip, fed across the top of the die. The stripper in this case is a flat plate screwed to the top of the die and contains a hole through which the punch passes. When the punch is on the upward return stroke, the plate prevents the material from lifting and strips it off the punch.

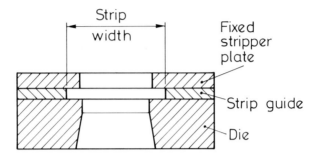

Fig. 3.8 Fixed stripper

When the work material is not in the form of strip and has to be loaded in the tool by hand, the spring-pad type of stripper is used, fig. 3.9. The stripper pad is set in advance of the punch, holding the material on the die face by means of springs and keeping it flat while the operation is being carried out. Stripper bolts keep the pad in position. When the punch is on the upward return stroke, the pad holds the material against the die and strips it off the punch.

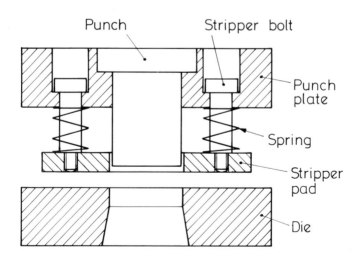

Fig. 3.9 Spring-pad stripper

Bending

Bending as described here refers to simple bending, confined to straight bends across the work material in one plane only. In bending operations, no metal cutting takes place: the previously cut material is placed between a punch and a die and force is applied to form the required bend.

Strip or sheet metal should, wherever possible, be bent in a direction across the grain of the material rather than along it. The direction of the grain is produced in the rolling process, and by bending across it there is less tendency for the material to crack. Keeping the bend radius as large as possible will also reduce the tendency for the material to crack.

Bends should not be positioned close to holes, as these can be pulled into an oval shape. It is generally accepted that the distance from the centre of the bend to the edge of a hole should be at least two-and-a-half times the thickness of the work material.

In bending operations, the length of the blank before bending has to be calculated. Any metal which is bent will stretch on the outside of the bends and be compressed on the inside. At some point between the inside and outside faces, the layers remain unaltered in length and this point is known as the neutral axis. For bends of radius more than twice the material thickness, the neutral axis may be assumed to lie at the centre of the material thickness, fig. 3.10(a). For sharper bends of radius less than twice the material thickness, the neutral axis shifts towards the inside face. In this case, the distance from the neutral axis to the inside face may be assumed to be 0.33 times the material thickness, fig. 3.10(b).

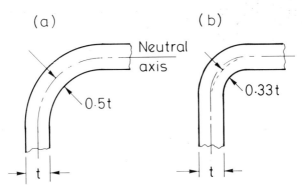

Fig. 3.10 Position of neutral axis

The length of the blank is determined by calculating the lengths of the flat portions either side of the radius plus the stretched out length of the bend radius (known as the bend allowance).

Example 1 Determine the blank length of the right-angled bracket shown in fig. 3.11(a).

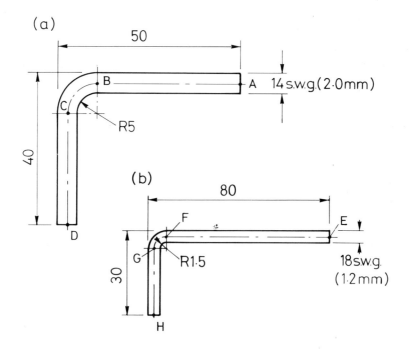

Fig. 3.11 Bending examples

Length AB = 50 mm − inside radius − material thickness
= 50 mm − 5 mm − 2 mm
= 43 mm

Length CD = 40 mm − 5 mm − 2 mm
= 33 mm

Since the radius is greater than twice the material thickness t, we can assume that the distance from the inside face to the neutral axis is $0.5t$.

∴ radius to neutral axis = 5 mm + (0.5 × 2 mm) = 6 mm

Since it is a 90° bend, length BC equals a quarter of the circumference of a circle of radius 6 mm

∴ length BC = $\dfrac{2\pi R}{4}$ = $\dfrac{\pi R}{2}$ = $\dfrac{6\pi \text{ mm}}{2}$ = 9.4 mm

∴ blank length = 43 mm + 33 mm + 9.4 mm = 85.4 mm

Example 2 Determine the blank length of the right-angled bracket shown in fig. 3.11(b).

Length EF = 80 mm − 1.5 mm − 1.2 mm = 77.3 mm

Length GH = 30 mm − 1.5 mm − 1.2 mm = 27.3 mm

Since the radius in this case is less than twice the material thickness t, we can assume that the distance from the inside face to the neutral axis is $0.33t$.

∴ radius to neutral axis = 1.5 mm + (0.33 × 1.2 mm) = 1.9 mm

∴ length FG = $\dfrac{\pi R}{2}$ = $\dfrac{1.9\pi \text{ mm}}{2}$ = 2.98 mm, say 3 mm

∴ blank length = 77.3 mm + 27.3 mm + 3 mm = 107.6 mm

Another factor which must be considered in bending operations is the amount of springback. Metal that has been bent retains some of its original elasticity and there is some elastic recovery after the punch has been removed. This is known as springback. In most cases this is overcome by overbending, i.e. bending the metal to a greater extent so that it will spring back to the required angle.

Die-sets

The punch is held in a punch plate and in its simplest form has a step as shown in fig. 3.12. The step prevents the punch from pulling out of the punch plate during operation.

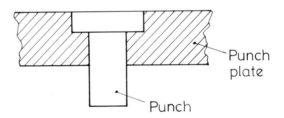

Fig. 3.12 Location of punch in punch plate

In order to ensure perfect alignment in the press, the punch plate and the die are secured in a die-set. Standard die-sets are available in steel or cast iron, and a typical example is shown in fig. 3.13. The top plate carries a spigot which is located and held in the press slide. The bolster contains slots for clamping to the press bedplate. The bolster has two guide pins on which the top plate slides up and down through ball bushes which reduce friction and ensure accurate location between the two parts.

The punch plate is fixed to the underside of the top plate and the die is fixed to the bolster, the punch and die being accurately aligned with each other.

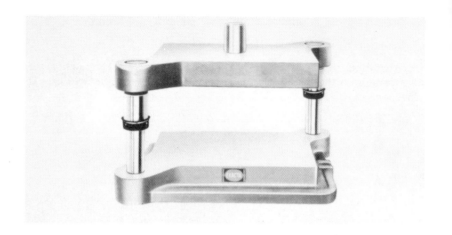

Fig. 3.13 Standard die-set

The die-set is mounted in the press as a complete self-contained assembly and can be removed and replaced as often as required in the knowledge that accurate alignment of punch and die is always maintained. Change-over times are low, since the complete die-set is removed and replaced by another die-set complete with its punches and dies for a different component.

3.3 Blanking, piercing, and bending operations

Specific objective To state the order of procedure for given examples of blanking, piercing, and bending.

Simple blanking
As previously stated, blanking is the production of an external shape from sheet metal. In its simplest form, this operation requires one punch and die.

For simple blanking from strip fed by hand, the press tool consists of a die on top of which are attached the strip guides and the stripper plate. The punch is held in the punch plate. The arrangement is shown in section in fig. 3.14.

To assist in setting up and for subsequent operation, stops are required. For setting up, a sliding stop is pushed in and the strip of work material is pushed against it by hand, fig. 3.15(a). The punch descends and blanks the first part, the blank falling through the die opening and out through the bedplate. On the upward stroke, the work material is stripped from the punch by the stripper plate. The sliding stop is then retracted and the work material is pushed up to the fixed stop locating in the hole produced in the blanking operation, fig. 3.15(b). This maintains a constant pitch between

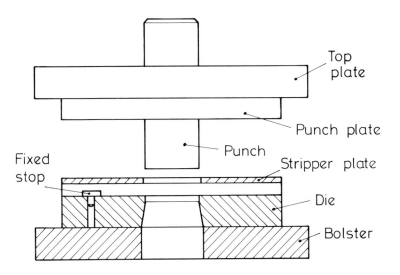

Fig. 3.14 Simple blanking tool

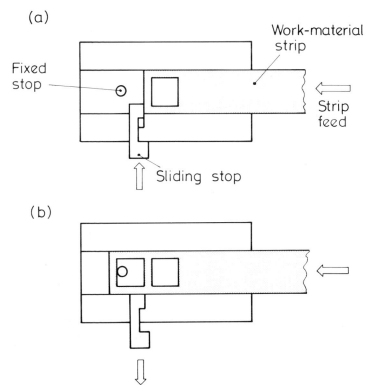

Fig. 3.15 Use of stops in simple blanking

blankings. The punch then descends and produces another blank; the punch is raised; the strip is moved forward against the fixed stop; and the operation is repeated. Thus a blank is produced each time the punch descends.

Blanking and piercing
Where the required workpiece is to have both an external and an internal shape – for example, a washer – the two operations, i.e. blanking and piercing, can be done by the same press tool. This type of press tool is known as a follow-on tool.

The principle is the same as for simple blanking but the punch plate has two punches fitted – one for blanking and one for piercing – and there are two holes of the required shape in the die, fig. 3.16. The work material is again in the form of strip, fed by hand.

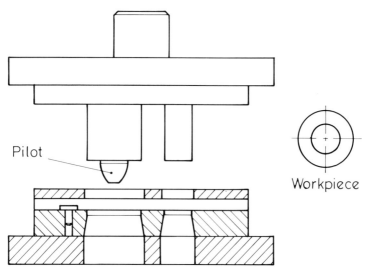

Fig. 3.16 Follow-on tool

For setting up, two sliding stops are required. The first sliding stop is pushed in and the strip of work material is pushed against it, fig. 3.17(a). The punches descend and the hole for the first workpiece is pierced. The punches are then raised. The first sliding stop is then retracted, the second sliding stop is pushed in, and the work material is pushed against it, fig. 3.17(b) – again to maintain a constant pitch. At this stage the pierced hole is now positioned under the blanking punch. The punches once more descend, the blanking punch producing a completed workpiece and at the same time the other punch piercing a hole. The second sliding stop is withdrawn and the work material is now moved forward against the fixed stop, fig. 3.17(c), again positioning the pierced hole under the blanking punch.

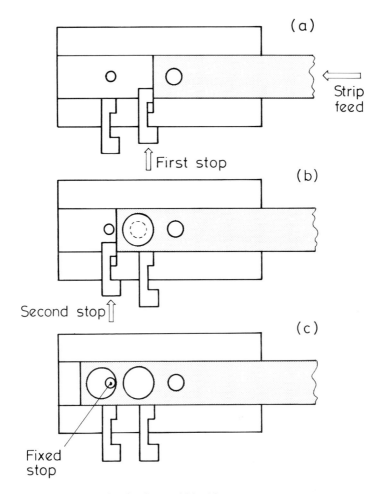

Fig. 3.17 Use of stops in piercing and blanking

The operation is then repeated, a completed workpiece being produced at each stroke of the press.

Greater accuracy of the inner and outer profiles of a workpiece can be obtained by fitting a pilot in the blanking punch (fig. 3.16). In this case the fixed stop is used for approximate positioning and is arranged so that the work material is drawn slightly away from it as the pilot engages in the pierced hole.

Bending
Two bending methods are commonly used, one known as vee bending and the other as side bending.

Vee-bending tools consist of a die in the shape of a vee block and a wedge-shaped punch, fig. 3.18. The metal to be bent is placed on top of the die – suitably located to ensure that the bend is in the correct position – and the punch is forced into the die. To allow for springback, the punch is made at an angle less than that required of the finished article. This is determined from experience – e.g. for low-carbon steel an angle of 88° is usually sufficient to allow the metal to spring back to 90°.

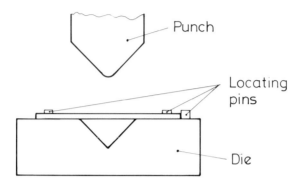

Fig. 3.18 Vee-bending tool

Side-bending tools are more complicated than those employed in vee bending but give a more accurate bend. The metal to be bent is placed on top of the die and is pushed against the guide block, which determines the length of the bent leg. Where the leg length is short, location pins can be used. As the punch plate descends, the pressure pad contacts the surface of the metal in advance of the punch and holds it against the die while the punch forms the bend. The guide plate prevents the punch from moving away from the work material during bending and helps the punch to iron the material against the side of the die, so preventing springback. The arrangement of this type of tool is shown in fig. 3.19.

3.4 Blanking layouts

Specific objective   To prepare blanking layouts which give maximum economic utilisation of material.

When parts are to be blanked from strip material, it is essential that the blank is arranged within the strip to gain the greatest economical use of the material by minimising the amount of scrap produced. The final layout will determine the width of strip, which in turn determines the general design and dimensions of the press tool.

The layout may be influenced by subsequent operations such as bending. In this case it is necessary to consider the direction of grain flow, as previously outlined.

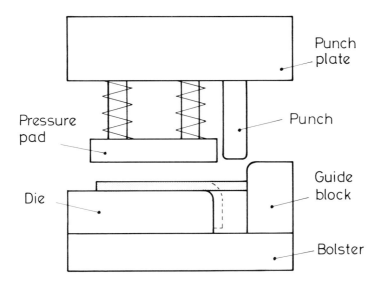

Fig. 3.19 Side-bending tool

It is also necessary to consider the minimum distance between blanks and between blanks and the edge of the strip – this distance must be large enough to support the strip during blanking. Insufficient distance results in a weakened strip which is subject to distortion or breakage, leading to misfeeding. The actual distance depends on a number of variables, but for our purpose a distance equal to the work-material thickness is acceptable.

The material utilisation can be calculated from the area of the part divided by the area of strip used in producing it, given as a percentage. The area of strip used equals the strip width multiplied by the feed distance. An economical layout should give at least a 75% material utilisation.

By virtue of their shape, some parts are simple to lay out whereas others are not so obvious.

Consider a 30 mm × 20 mm blank to be produced from 14 s.w.g. (2.0 mm) material. This would simply be laid out in a straight line as shown in fig. 3.20.

Since the work-material thickness is 2 mm, the distance between blanks and between blanks and the edges will be assumed to be 2 mm; therefore

strip width = 30 mm + 2 mm + 2 mm = 34 mm

and the distance the strip must feed at each stroke of the press in order to produce one blank is

20 mm + 2 mm = 22 mm

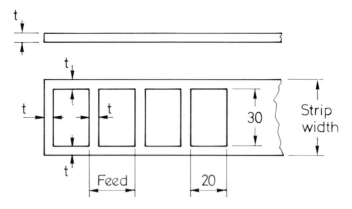

Fig. 3.20 Blank layout

$$\text{Material utilisation} = \frac{\text{area of part}}{\text{area of strip used}} \times 100\%$$

$$= \frac{\text{area of part}}{\text{strip width} \times \text{feed distance}} \times 100\%$$

$$= \frac{30 \text{ mm} \times 20 \text{ mm}}{34 \text{ mm} \times 22 \text{ mm}} \times 100\% = 80\%$$

Now consider the blank shown in fig. 3.21, to be produced from 19 s.w.g. (1.0 mm) material. The simple layout would be as shown in fig. 3.22(a). This gives a strip width of 32 mm and a feed of 31 mm, allowing 1 mm between blanks and between blanks and the edges of the strip. Thus the area of strip used per blank would be 32 mm × 31 mm = 992 mm².

Area of blank = 675 mm²

$$\therefore \text{ material utilisation} = \frac{675 \text{ mm}^2}{992 \text{ mm}^2} \times 100\% = 68\%$$

An alternative layout with the blanks turned through 45° is shown in fig. 3.22(b). In this case the strip width is 45 mm and the feed 23 mm. Thus the area of strip used per blank would be 45 mm × 23 mm = 1035 mm².

Area of blank = 675 mm²

$$\therefore \text{ material utilisation} = \frac{675 \text{ mm}^2}{1035 \text{ mm}^2} \times 100\% = 65\%$$

A further alternative layout is shown in fig. 3.22(c), with the blanks in an alternating pattern. This gives a strip width of 48 mm and a feed of 32 mm, which in this case produces two blanks. Therefore the feed per

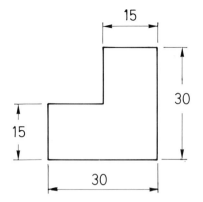

Fig. 3.21 Blank

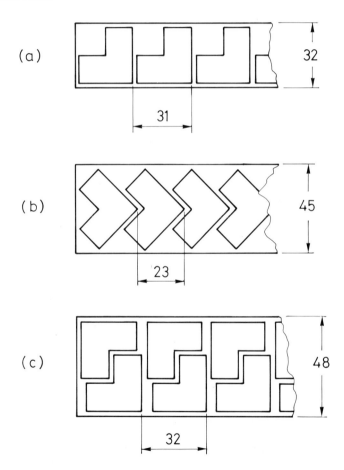

Fig. 3.22 Blank layout

blank is half this amount, i.e. 16 mm. Thus the area of strip used per blank would be 48 mm × 16 mm = 768 mm².

Area of blank = 675 mm²

$$\therefore \text{ material utilisation} = \frac{675 \text{ mm}^2}{768 \text{ mm}^2} \times 100\% = 88\%$$

This gives the most economical use of the material and would therefore be the obvious choice. However, with this layout the strip would have to be worked in two passes. On the first pass the bottom row would be blanked; the strip would then be turned round and passed through again for the other row to be blanked.

Exercises on chapter 3
1   What is the reason for inclining a gap-frame power press?
2   State a restriction in the use of a fly press.
3   How is protection against overload provided for in a hydraulic power press?
4   Why is the clearance between punch and die important in a press tool?
5   Name three types of presswork operation.
6   What is the purpose of a die-set?
7   Why is some form of stripper necessary in a press tool?
8   Name two sources of power for power presses.
9   How is clearance between punch and die calculated?
10   Name three ways in which power presses may be guarded.
11   What is 'springback' and how can it be overcome?
12   Why is the blanking layout important in presswork?

4 Plastics moulding

General objective The student understands the basic plastics-forming methods of compression moulding, transfer moulding, and injection moulding.

Most plastics used today are man-made and are described as 'synthetic' materials, i.e. they are made by a process of building up from simple chemical substances.

With a few exceptions, plastics are compounds of carbon with one or more of the five elements hydrogen, oxygen, nitrogen, chlorine, and fluorine. These compounds form a diverse group of different materials, each with its own characteristic properties and uses. All plastics materials are based on large molecules which are made by joining together large numbers of smaller molecules. The small molecules, known as monomers, are derived from natural gas and crude oil and are subjected to suitable conditions to join up and form long-chain-molecular products known as polymers. The process of joining the molecules together is known as polymerisation, and the names of the plastics made in this way frequently contain the prefix 'poly'. For example, the monomer ethylene is polymerised to form very-long-chain molecules of the polymer polyethylene (also called 'polythene').

Plastics are mostly solid and stable at ordinary temperatures, and at some stage of their manufacture they are 'plastic', i.e. soft and capable of being shaped. The shaping process is done by the application of heat and pressure, and it is the behaviour of the material when heated that distinguishes between the two classes of plastics: thermoplastics and thermosetting plastics.

Plastics made up of molecules arranged in long chain-like structures which are separate from each other soften when heated and become solid again when cooled. By further heating and cooling, the material can be made to take a different shape, and this process can be repeated again and again. Plastics having this property are known as thermoplastics and examples include polyethylene (polythene), p.v.c., polystyrene, ABS, acrylics, polypropylene, nylon, and p.t.f.e.

Other plastics, although they soften when heated the first time and can be shaped, become stiff and hard on further heating and cannot be softened again. During the heating process, a chemical reaction takes place which cross-links the long chain-like structures, thus joining them firmly and permanently together – a process known as curing. Plastics of this type are known as thermosetting plastics, or thermosets, and examples include phenolics ('Bakelite') and aminos (urea formaldehyde and melamine formaldehyde).

As already stated, the shaping of plastics material is achieved by the application of heat and pressure. There are a great many ways in which this can be done, depending on the nature of the polymer, the type and size of product, and the quantity and dimensional accuracy required. The methods to be described here are moulding by the compression, transfer, and injection processes.

Before dealing with the moulding process itself, however, it is necessary to consider the polymer material, which is usually unsuitable for moulding until mixed with other ingredients called additives.

4.1 Forms of supply

Specific objective To describe the forms of supply of the materials used for compression, transfer, and injection moulding.

Plastics materials for use in moulding are normally in the form of powders or small chips known as granules, or as preforms.

In their pure unmodified state, polymer materials resulting from industrial polymerisation are in most cases unsuitable for processing into finished articles. Before being moulded, they have to be mixed with other ingredients – known as additives – in order to modify or eliminate undesirable properties and to develop their useful characteristics.

Additives can be incorporated into the monomer before polymerisation, during the polymerisation reaction, or with the polymer itself. Some additives modify the properties of the polymer by physical means, while some achieve their effect by chemical reactions and are used not only to influence the properties of the finished product but also to improve processing characteristics. Some of the more common additives are as follows.

a) *Plasticisers* – added particularly to p.v.c. to give greater flexibility and make it easier to form.
b) *Stabilisers* – added, again particularly to p.v.c., to prevent decomposition of the polymer at temperatures encountered during normal processing.
c) *Lubricants* – widely used to facilitate the processing of a variety of polymers, by reducing the forces between molecules and by reducing adhesion of the polymer to hot metal surfaces during processing.
d) *Antioxidants* – added to prevent oxidative degradation, i.e. gradual breakdown in the presence of oxygen, which most polymers are subject to at the elevated temperatures necessary for processing and at atmospheric temperatures over a period of time.
e) *Fillers* – added to improve physical properties or in some cases to give a cheaper product by acting as extenders. Fillers used include wood flour, cork dust, asbestos, carbon black, chalk, and chopped glass fibre.

f) *Ultra-violet absorbers* – added to protect polymers from the adverse affect of exposure to ultra-violet radiation, the main source of which is sunlight.
g) *Flame retardants* – added as most polymers are flammable to a greater or lesser extent.
h) *Colourants* – added for a number of reasons:
   i) to give greater product appeal and make the product more saleable;
   ii) as a means of identification (e.g. cable insulation);
   iii) to make the product more readily visible (e.g. garments for roadmen and motor cyclists);
   iv) to simulate a natural or traditional product (e.g. leather luggage).

Colourants may be added as dyes or, more commonly, as pigments and are available in a vast range of colours.

4.2 Moulding processes

Specific objective To describe with the aid of sketches the basic steps of compression, transfer, and injection moulding.

Metals have a definite melting point and in general tend to be free-flowing in a molten state. Polymers, on the other hand, have no definite melting point but are softened by the application of heat, which renders them 'plastic'. In this state they may be considered as very viscous fluids and, as a result, high pressures are required for moulding.

The viscosity of a polymer is reduced by the application of heat, but there is an upper temperature limit at which the polymer begins to break down in some way. This breakdown is known as degradation. All polymers are bad conductors of heat and are therefore susceptible to overheating. If a polymer is subjected to excessive temperature or to prolonged periods in the mould, degradation will occur.

There is also a lower limit of temperature below which the polymer will not be soft enough to flow into the mould. The temperature for moulding must be between the upper and lower limits and will directly affect the viscosity of the polymer.

All the moulding processes require three stages:

i) application of heat to soften the moulding material,
ii) forming to the required shape in a mould,
iii) removal of heat.

The three moulding techniques to be discussed essentially differ only in the way the moulding material is heated and delivered to the mould.

Compression moulding

Compression moulding is used for thermosetting plastics. The process is carried out in a hydraulic press with heated platens. The two halves of the

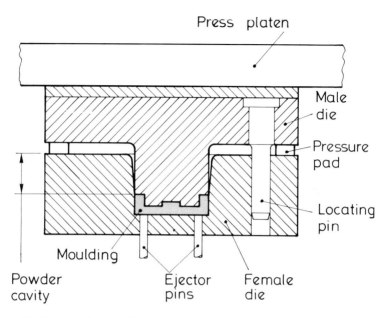

Fig. 4.1 Compression mould

moulding tool consist of a male and a female die, to give a cavity of the required finished shape of the product, and are attached to the platens of the press, fig. 4.1. Depending on the size and shape of the product and on the quantity required, the moulding tool may contain a single cavity or a number of cavities, when it is known as a multi-cavity mould. The mould cavity is designed with an allowance for shrinkage of the moulding material and a draft angle of at least 1° to allow for the escape of gases and easy removal of the product after moulding. Moulding tools are manufactured from tool steel, hardened and tempered to give strength, toughness, and a good hard-wearing surface. The male and female dies are highly polished and, once the tool has been proved, these surfaces are chromium-plated (around 0.005 mm thickness) to give a high surface gloss to the product and to facilitate its removal and protect the surfaces from the corrosive effects of the moulding material.

Moulding materials are normally in the form of loose powder or granules. These materials have a high bulk factor, i.e. the volume of the loose material is much greater than that of the finished product. The bulk factor is around 2.5:1 for granulated materials and can be as high as 4:1 for fine powder. To allow for this, a powder cavity is built into the female die attached to the bottom press platen. To prevent an excess of loose material being loaded into the tool, each charge is weighed, either on scales or by some automatic method.

Alternatively, loose powder material can be compressed to form a small pellet, or preform, of a size and shape to suit the mould cavity. This is done cold, so that no curing takes place, and is carried out in a special preforming or pelleting machine. The preform is easily handled and gives a consistent mass of charge. Depending on the material, these preforms may be preheated to around 85°C in a high-frequency oven – this reduces the cycle time, since the preform is partially heated before it is loaded in the tool and requires less pressure in moulding.

The typical sequence of operations for compression moulding is as follows.

i) Load moulding material – as loose powder, granules, or a heated preform – into the heated die cavity. Moulding temperatures vary, e.g. between 135°C and 155°C for urea powders and between 140°C and 160°C for melamine materials.
ii) Close the split mould between the press platens. The pressure is around 30 N/mm^2 to 60 N/mm^2 (lower for preheated pellets). The combined effect of the moulding temperature and pressure causes the moulding material to soften and flow into the mould cavity. Further exposure to the moulding temperature causes the irreversible chemical reaction of cross-linking or curing. The curing time depends on the wall thickness, mass, moulding material, and moulding temperature; for example, a 3 mm section of urea will cure in around 30 s at 145°C and the same section in melamine in up to 2 min at 150°C.
iii) Open the split mould.
iv) Eject the product from the mould. This may be done by hand or automatically, according to the complexity of the tool or product. Since the material is thermosetting and has cured, there is no need to wait for the moulding to cool and therefore it can be removed immediately while it is still hot.
v) Blow out the tool to remove any particles left behind by the previous moulding.
vi) Lubricate the tool to assist the release of the next moulding.
vii) Repeat the process.

Any material which escapes during the moulding process results in a feather edge on the moulding known as a flash. This can be removed during the curing cycle of the next moulding.

Compression moulding is used to produce a wide range of products, including electrical and domestic fittings, toilet seats and covers, bottle tops and various closures, and tableware.

Transfer moulding

Transfer moulding is used for thermosetting plastics. This process is similar to compression moulding except that the plasticising and moulding functions are carried out separately. The moulding material is heated until plastic in a transfer pot from which it is pushed by a plunger through a series of runners into the heated split mould where it cures, fig. 4.2. The two halves of the split mould are attached to the heated platens of a hydraulic press in the same way as for compression moulding.

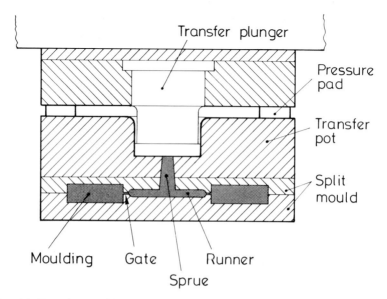

Fig. 4.2 Transfer mould

The split mould is closed and the moulding material in the form of powder, granules, or a heated preform is placed in the transfer pot and pressure is applied. The pressure on the area of the transfer pot is greater than that in the compression-moulding process but, as the moulding material is plastic when it enters the mould cavity, the pressure within the cavity is much less. As a result of this, the process is suited to the production of parts incorporating small metal inserts. Intricate parts and those having variations of section thickness can be produced to advantage by this method. Cure times are less and greater accuracy is achieved than with compression moulding.

The main limitation of the process is the loss of material in the sprue, runners, and gates – as thermosetting materials cure during moulding, this cannot be reused. Moulding tools are usually more complex and therefore more costly than compression-moulding tools.

Typical products produced by this process are motor-car distributor caps and domestic electric plugs.

Injection moulding
Injection moulding may be used for either thermoplastics or thermosetting materials, but is most widely used for thermoplastics.

The major advantages of this moulding process are its high production rate, high degree of dimensional accuracy, and its suitability for a wide range of products.

The moulding material is fed by gravity from a hopper to a cylindrical heating chamber where it is rendered plastic and then injected into a closed mould under pressure. The moulding solidifies in the mould. On solidification, the mould is opened and the moulding is ejected.

The type of injection-moulding machine most widely used is the horizontal type, fig. 4.3. A hopper at the opposite end of the machine from the mould is charged with moulding material in the form of powder or granules which is fed by gravity to the heating chamber. Electric band heaters are attached to the outer casing of the heating chamber, inside which is an extruder-type screw similar to that of a domestic mincing machine. As the screw rotates, it carries the material to the front of the heating chamber. The band heaters and the frictional forces developed in the material by the rotating screw result in the material becoming plastic as it passes along.

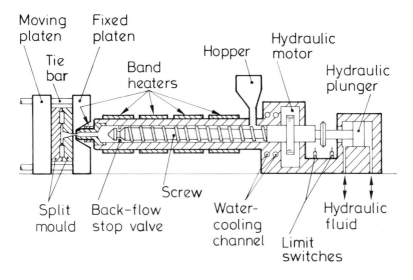

Fig. 4.3 Injection-moulding machine

When the plasticised material builds up in front of the screw, the screw moves axially backwards and its rotation is stopped. The amount of material to be injected into the mould – known as the shot size – is controlled by stopping the screw rotation at a predetermined position. At this stage, the mould, filled by the previous shot, is opened and the moulding is ejected. The mould is then closed and the stationary screw is moved axially towards the mould, pushing the plasticised material into the mould cavity under pressure. The screw then rotates, feeding more moulding material along the heating chamber to become plasticised, the material being continuously replaced from the hopper. The screw then moves axially backwards due to the build-up, and the sequence recommences. The heating-chamber temperature varies between 120°C and 260°C, depending on the type of moulding material and the shot size.

Moulding tools are of the same materials and finish as described for compression-moulding tools but without, of course, the need for a powder cavity. They are generally more expensive than compression-moulding tools but, because of the higher production rates of injection moulding, it is possible to use smaller, i.e. single-cavity, and hence cheaper tools and maintain the same number of mouldings per hour as would be possible with a compression press.

The use of thermoplastics material requires that the mould be maintained at a constant temperature – usually around 75°C to 95°C – to cool and solidify the material within the mould before the moulding can be ejected. This is achieved by circulating water through the mould and makes the process much faster than compression moulding. Although material is used in the sprue and runners, material wastage is low since it can be reused.

Injection moulding of thermosetting materials is achieved in the same way as for thermoplastic materials except that the temperatures used are more critical. The temperature of the heating chamber is important – to avoid the moulding material curing before it enters and fills the mould cavity – and is between 95°C and 105°C for urea materials and between 100°C and 110°C for melamine materials. Moulding-tool temperature is also important, to ensure correct curing of the moulding material. Tool temperatures between 135°C and 145°C for urea materials and between 145°C and 155°C for melamine materials are suitable for sections over 3 mm and are increased by 10°C for sections below 3 mm. As the moulding material enters the mould at very near to the curing temperature, cycle times are low.

The range of injection-moulded components is vast and includes toys, e.g. model kits; houseware, e.g. buckets, bowls, and washing-machine parts; and car components.

4.3 Inserts

Specific objective   To explain where inserts may be used to advantage.

The primary purpose of inserts is to strengthen relatively weak plastics materials in order to facilitate the joining of mouldings or the mounting of other parts to the moulding. Inserts may also be used for electrical purposes, to provide a conductor in an otherwise insulating material.

Inserts are usually made from steel or brass and may be moulded-in during the moulding process or introduced into a plain hole produced during moulding or by a subsequent drilling operation. This latter type may be installed in the plain hole by using heat (thermal or ultrasonic), by pushing in, or by a self-tapping arrangement. Installation using heat is normally confined to the thermoplastic materials. Whichever type of insert is used, it is essential that an adequate means of anchorage is provided to prevent movement or pulling out of the insert in use.

Use of inserts for electrical purposes can be illustrated by a motor-vehicle distributor. The distributor cap has moulded-in inserts as shown in the cut-away view in fig. 4.4. Examples of inserts for fixing or mounting purposes, again moulded-in, can be seen in the 13 A plug top and the electrical connection box shown in fig. 4.5.

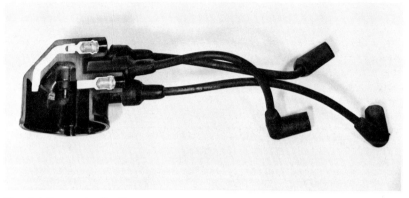

Fig. 4.4 Inserts in distributor cap

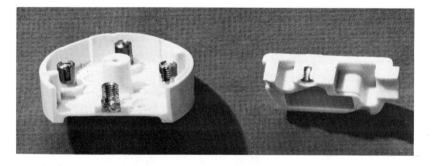

Fig. 4.5 Inserts in electrical components

A wide variety of threaded inserts, both internal and external, are available with a variety of exterior arrangements for inserting and anchoring them in a premoulded or predrilled hole. Figure 4.6 shows internally threaded inserts with a 'barb' form of exterior. These may be installed by pushing in or, in the case in the illustration, by ultrasonic installation where ultrasonic vibration causes instant localised melting in the clear acrylic knob. Figure 4.7 shows externally threaded inserts – the larger thread form is tapered and contains a slot to give a self-tapping action. These self-tapping inserts are installed by hand or power tools to give a finished product as shown.

Fig. 4.6 Internal-threaded inserts

Fig. 4.7 External-threaded inserts

4.4 Advantages and limitations of moulding processes

Specific objectives To list the principal advantages and limitations of the compression-, transfer-, and injection-moulding processes and to select the process appropriate for a given component.

Advantages and limitations of the moulding processes are summarised in Table 4.1.

Table 4.1 Advantages and limitations of moulding processes

	Moulding process		
	Compression	Transfer	Injection
Capital machine cost	Medium May require preheat ovens and pelleting machines.	Medium	High
Moulding-tool cost	Low	Medium	High
Output rate	Low	Medium	High
Dimensional accuracy	Low	High	High
Moulding material	Thermosets	Thermosets	Mainly thermoplastics
Finishing required	Flash removal	Removal of flash, sprue, and runners	Removal of sprue and runners
Inserts moulded in?	Not usual	Yes	Yes
Waste material	None	Sprue and runners cannot be reused	With thermoplastics, sprue and runners can be reused

To select the appropriate moulding process to be used for a given component, a number of factors may be considered, some of which may automatically exclude a particular process. Among these factors, referred to in Table 4.1, are

a) the type of material to be moulded,
b) the dimensional accuracy required,
c) the required output.

For example, a small quantity of mouldings to be made from a thermosetting material and not requiring a high degree of accuracy would favour the compression-moulding process; larger quantities of higher accuracy from a thermosetting material would favour transfer moulding; while high-volume production from thermoplastics material would be carried out using the injection-moulding process.

4.5 Safety in plastics moulding

Specific objective To identify sources of danger in using plastics-moulding machines and to outline the necessary precautions.

Any machine with moving parts is a potential hazard and as such must be properly guarded. Moulding machines used with plastics are basically presses using high forces to lock the dies in position while the moulding operation is carried out. Adequate guarding must therefore be provided and used, to prevent the operator coming in contact with moving parts and so eliminate the possibility of trapped fingers or hands.

Heat is an essential part of any moulding process. Moulding materials have a high heat capacity and in their hot plastic state will stick on contact with the skin and are difficult to remove. Protection can be afforded by the use of suitable industrial gloves. Some of the materials used can cause dermatitis, which can be prevented by the use of gloves.

Harmful gases and vapours are given off by some plastics materials, so the moulding machine must be fitted with adequate extraction equipment.

Exercises on chapter 4
1 What is the purpose of using fillers in plastics materials?
2 Can the waste material after moulding thermoplastic and thermosetting materials be reused? Give the reason for your answer.
3 What is meant by 'polymerisation'?
4 What is the purpose of preforming moulding powder?
5 Why is it undesirable to process thermoplastics materials by compression moulding?
6 State an advantage of transfer moulding over compression moulding.
7 What is the difference between thermoplastics and thermosetting plastics materials?
8 Name four types of thermoplastics materials.
9 What is the purpose of additives in a polymer?
10 Why is a powder cavity necessary in a mould for compression moulding?

5 Measurement

General objective The student knows the procedures for measurement of dimension and form and the importance of workshop standards.

Standards are necessary in industry in order to ensure uniformity and to establish minimum requirements of quality and accuracy. The adoption of standards eliminates the national waste of time and material involved in the production of an unnecessary variety of patterns and sizes of articles for one and the same purpose. Standards are desirable not only in the manufacture of articles but also for the instruments used to ensure the accuracy of these articles.

The word 'standard' can refer to a physical standard such as length or to a standard specification such as a paper standard. In the United Kingdom, physical standards are maintained by the National Physical Laboratory (NPL). As the nation's standards laboratory, NPL provides the measurement standards and calibration facilities necessary to ensure that measurements in the UK may be carried out on a common basis and to the required accuracy. The national primary standards, which constitute the basis of measurement in the UK, are maintained at NPL in strict accordance with internationally agreed recommendations. They are based on the International System of Units (SI).

Within the UK, the national primary standards are used to calibrate secondary standards and measuring equipment manufactured and used in industry. This calibration service is provided by NPL or through the approved laboratories of the NPL-based British Calibration Service (BCS). BCS is a national service which, for particular measurements, specially approves laboratories which are then authorised to issue official certificates for such measurements. These laboratories are located in industry, educational institutions, and government establishments. A British Calibration Service certificate issued by an approved laboratory indicates that measurements are traceable to national standards and provides a high degree of assurance of the correctness of the calibration of the equipment or instrument identified on the certificate.

The preparation of standard specifications in the UK is the responsibility of the British Standards Institution (BSI), whose main function is to draw up and promote the adoption of voluntary standards and codes of good practice by agreement among all interested parties, i.e. manufacturers, users, etc. BSI plays a large and active part in the work of the International Organisation for Standardisation (ISO), which is responsible for the preparation of international recommendations.

The following sections will deal with the measurement of dimension and of form – such as straightness, flatness, roundness, and squareness.

5.1 Length

Specific objective To state the principles of use and care of workshop standards for length.

The universal standard of length is the metre, and the definition of this in terms of wavelengths of light was agreed by all countries in 1960. The metre is defined as 1 650 763.73 wavelengths of the orange radiation of the krypton-86 isotope in vacuo. At the same time the yard was defined as 0.9144 m, which gives an exact conversion of 1 inch = 25.4 mm. The great advantage of a wavelength standard is that it is constant, unlike material-length standards where small changes in the length of metal bars can occur during periods of time. Also, a wavelength standard is directly transferable to a material standard in the form of an end gauge, e.g. gauge blocks, and to a line standard, i.e. a measuring scale such as a vernier scale, to a high degree of accuracy.

End standards of length used in the workshop are of two types: gauge blocks and length bars. These are calibrated and intended for use at 20°C.

Gauge blocks

Gauge blocks are made from a wear-resisting material – hardened and stabilised high-grade steel or tungsten carbide – and their dimensions and accuracy are covered by BS 4311:part 1:1968, 'Specification for metric gauge blocks. Gauge blocks'.

Two uses for gauge blocks are recognised: (i) general use for precise measurement where accurate work sizes are required, (ii) as standards of length used with high-magnification comparators to establish the sizes of gauge blocks in general use.

Each gauge block is of rectangular section with the measuring faces finished by precision lapping to the required distance apart – known as the gauge length, fig. 5.1 – within the tolerances of length, flatness, and paral-

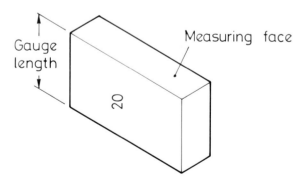

Fig. 5.1 Gauge block

lelism set out in the standard. The measuring faces are of such a high degree of surface finish and flatness that gauge blocks will readily wring to each other, i.e. they will adhere when pressed and slid together. Thus a set of selected-size gauge blocks can be combined to give a very wide range of sizes, usually in steps of 0.001 mm.

Standard sets are available with differing numbers of pieces. A typical set is shown in fig. 5.2. These sets are identified by a number indicating the number of pieces prefixed by the letter M, to indicate metric sizes, and followed by the number 1 or 2. This latter number refers to a 1 mm- or 2 mm-based series, this being the base gauge length of the smaller blocks. For example, an 88 piece set to a 1 mm base is designated M88/1 and contains the following sizes:

Size (mm)	Increment (mm)	Number of pieces
1.0005	–	1
1.001 to 1.009	0.001	9
1.01 to 1.49	0.01	49
0.5 to 9.5	0.5	19
10 to 100	10	10
	Total	88 pieces

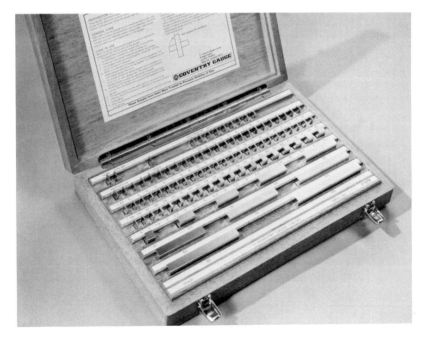

Fig. 5.2 Set of gauge blocks

Table 5.1 Tolerances on flatness, parallelism, and length of gauge blocks (BS 4311:part 1:1968)

Size of gauge block		Tolerances (unit = 0.01 μm (0.000 01 mm))											
		Calibration grades and grade 00				Grade 0			Grade I			Grade II		
Up to and				Gauge length L										
				Calibration										
Over	including	Flatness	Parallelism	grade	Grade 00	Flatness	Parallelism	Gauge length L	Flatness	Parallelism	Gauge length L	Flatness	Parallelism	Gauge length L
mm	mm													
—	20	5	5	±25	±5	10	10	±10	15	20	+20 / −15	25	35	+50 / −25
20	60	5	8	±25	±8	10	10	±15	15	20	+30 / −20	25	35	+80 / −50
60	80	5	10	±50	±12	10	15	±20	15	25	+50 / −25	25	35	+120 / −75
80	100	5	10	±50	±15	10	15	±25	15	25	+60 / −30	25	35	+140 / −100

An 88 piece set to a 2 mm base is designated M88/2 and contains the following sizes:

Size (mm)	Increment (mm)	Number of pieces
1.0005	–	1
2.001 to 2.009	0.001	9
2.01 to 2.49	0.01	49
0.5 to 9.5	0.5	19
10 to 100	10	10
		Total 88 pieces

The 2 mm-based series are recommended as they are less likely to suffer deterioration in flatness than the thinner 1 mm blocks.

The recommended sets in BS 4311 are M32/1, M41/1, M47/1, M88/1, M112/1, M32/2, M46/2, and M88/2.

Five grades of accuracy are provided for in the British Standard: grades 0, I, II, 00, and calibration. The tolerances on flatness, parallelism, and length for each grade are shown in Table 5.1.

The choice of grade is solely dependent on the application. Grade 0 is used for gauge inspection and high-precision work. Grade I is most commonly used in the production of components, tools, and gauges. Grade II is used for preliminary setting up of work and comparatively rough checking where work tolerances are relatively large. Grade 00 is used only where there is a need for the highest accuracy without referring to a calibration chart, i.e. where it is assumed that each block is true to its nominal size. This grade has the same tolerance for flatness and parallelism as the calibration grade, but the gauge-length tolerances are much smaller.

Calibration-grade gauges should not be used for general inspection work: they are intended for calibrating other gauge blocks. This means that the actual gauge length is known, and this is obtained by referring to a calibration chart of the set, i.e. a chart showing the actual size of each block in the set. For this reason, relatively large tolerances on gauge length can be allowed, but calibration-grade gauges are required to have a high degree of accuracy of flatness and parallelism.

To build up a size combination, the smallest possible number of gauge blocks should be used. This can be done by taking care of the micrometres (0.001 mm) first, followed by hundredths (0.01 mm), tenths (0.1 mm), and whole millimetres; for example, to determine the gauge blocks required for a size of 78.748 mm using the M88/2 set previously listed:

```
             78.748
  Subtract    2.008   . . . .  1st gauge block
             ──────
             76.740
  Subtract    2.24    . . . .  2nd gauge block
             ──────
             74.50             [carried forward]
```

```
         74.50           [brought forward]
Subtract  4.50    . . . . 3rd gauge block
         ─────
         70.00    . . . . 4th gauge block
```

The 2.24 mm second gauge block is used as it conveniently leaves a 0.5 increment for the third gauge block.

In some instances protector gauge blocks, usually of 2 mm gauge length, are supplied with a set, while in other instances they have to be ordered separately. These are available in pairs, are marked with the letter P, and are placed one at each end of a build-up to prevent wear on the gauge blocks in the set. If wear takes place on the protector gauge blocks, then only these need be replaced. Allowance for these, if used, must be made in calculating the build-up.

Having established the sizes of gauge blocks required, the size combination can be built up. Select the required gauge blocks from the case and close the lid. It is important that the lid is kept closed at all times when not in actual use, to protect the gauge blocks from dust, dirt, and moisture.

Clean the measuring faces of each gauge block with a clean chamois leather or a soft linen cloth and examine them for damage. Never attempt to use a gauge block which is damaged, as this will lead to damage of other gauge blocks. Damage is likely to occur on the edges through the gauge block being knocked or dropped, and, in the event of damage, it is preferable to return the gauge block to the manufacturer for the surface to be restored.

When two gauge blocks are pressed and slid together on their measuring surfaces, they will adhere and are said to 'wring' together. They will only do this if the measuring surfaces are clean, flat, and free from damage.

Wring two gauge blocks by placing one on top of the other as shown in fig. 5.3 and sliding them into position with a rotary movement. Repeat this for all gauge blocks in the size combination, starting with the largest gauge blocks. Never wring gauge blocks together while holding them above an open case, as they could be accidentally dropped and cause damage.

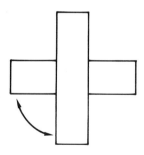

Fig. 5.3 Position for wringing two gauge blocks

Do not finger the measuring faces, as this leads to the risk of tarnishing, and avoid unnecessary handling as this can lead to an increase in temperature and hence dimension.

Immediately after use, slide the gauge blocks apart, clean each one carefully, replace in the case, and close the lid. Do not break the wringing joint but slide the blocks apart, and never leave the gauge blocks wrung together for any length of time.

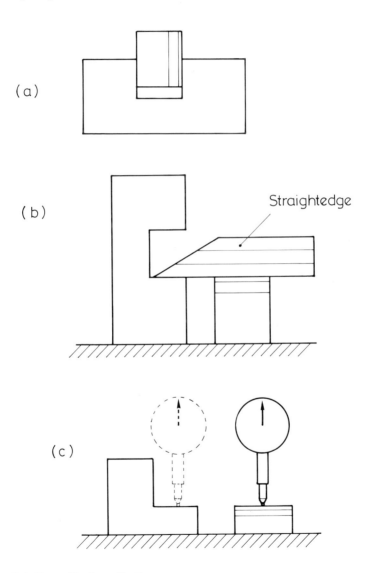

Fig. 5.4 Gauge-block applications

Gauge blocks, separately or in combination, may be used for direct measurement as shown in fig. 5.4(a) or for comparative measurement using a knife-edged straightedge or a dial indicator, figs 5.4(b) and (c). The precise size of the gauge-block combination is usually arrived at by trial and error. The gauge-block combination is built up until the top of the work and the gauge blocks are the same height, this being so when no light is visible under the knife edge of the straightedge when it is placed simultaneously on both surfaces, fig. 5.4(b), and viewed against a well illuminated background. The same principle is employed using a dial indicator, the work and the gauge-block combination being the same height when the same reading is obtained on the dial indicator from each surface, fig. 5.4(c).

Gauge blocks are also widely used in conjunction with sine bars and with calibrated steel balls and rollers, as well as for checking straightness and squareness. These will be dealt with later in the chapter. They are also used in conjunction with a range of accessories.

Gauge-block accessories The use of gauge blocks for measuring can be extended by the use of a range of accessories. These are covered by BS 4311:part 2:1977, 'Specification for metric gauge blocks. Accessories'. A typical set of gauge-block accessories is shown in fig. 5.5 and consists of

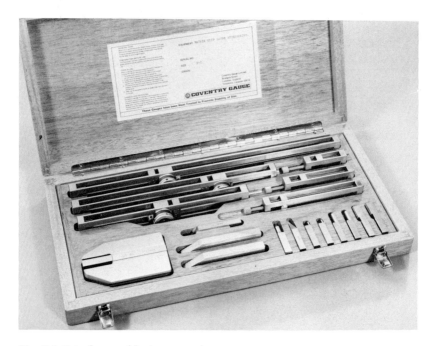

Fig. 5.5 Set of gauge-block accessories

i) two pairs of jaws, type A and type B, which when combined with gauge blocks form an external or internal caliper, figs 5.6(a) and (b);
ii) a centre point and a scriber, used in combination with gauge blocks to scribe arcs of precise radius, fig. 5.6(c);
iii) a robust base for converting a gauge-block combination together with the scriber into a height gauge, fig. 5.6(d);
iv) a knife-edged straightedge;
v) holders for supporting the various combinations when in use.

The accessories other than the holders are made from high-quality steel, hardened and stabilised, with their wringing faces precision-lapped to give flatness, parallelism, and surface finish to the same degree as gauge blocks, to which they are wrung to assemble any combination. Accessories can be purchased in sets as shown or as individual items.

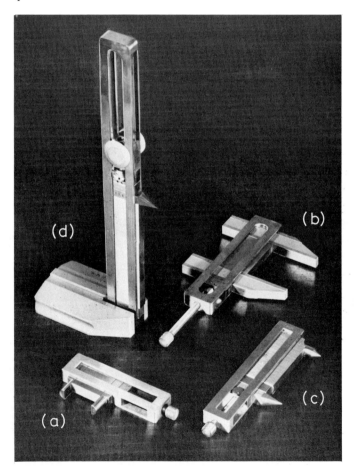

Fig. 5.6 Assembled gauge blocks and accessories

Length bars

Gauge-block combinations are suitable for providing end standards up to about 150 mm, but above this size they are difficult to handle. Length bars can be wrung together to form combination lengths well in excess of the range of gauge blocks.

Length bars are of round section, 22 mm diameter, and are made of the same high-quality steel as gauge blocks. The gauge faces are hardened and stabilised and finished by precision lapping to the required length, flatness, and parallelism set out in BS 5317:1976, 'Metric length bars and their accessories'. Length bars are available as individual gauges in lengths from 10 mm to 1200 mm or as sets supplied in a wooden case. Figure 5.7 shows a set of length bars and a set of length-bar accessories. This set contains nine bars of lengths 10, 20, 40, 60, 80, 100, 200, 300, and 400 mm.

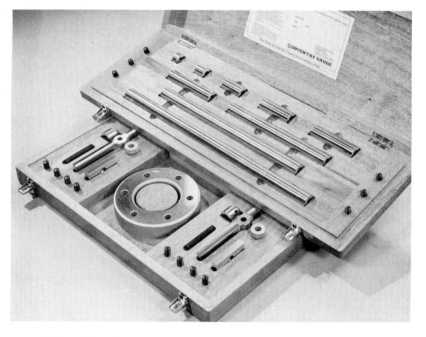

Fig. 5.7 Set of length bars and length-bar accessories

Four grades of accuracy are provided for in BS 5317 – reference, calibration, grade 1, and grade 2. The tolerances on flatness, parallelism, and length for each grade are shown in Table 5.2.

Reference-grade bars are intended for use as reference standards and are of the highest accuracy. They should be used only in a 'standards room', temperature-controlled at 20°C.

Calibration-grade bars are intended for use in the calibration of length-measuring standards and again should only be used in a 'standards room' at 20°C.

Table 5.2 Tolerances on flatness, parallelism, and length of length bars (BS 5317:1976)

Nominal length	Tolerances (unit = μm = 0.001 mm)									
	Reference and calibration grade				Grade 1			Grade 2		
	Flatness	Parallelism	Length		Flatness	Parallelism	Length	Flatness	Parallelism	Length
			Reference grade	Calibration grade						
up to 25	0.08	0.08	±0.08	±0.15	0.15	0.16	+0.40 / −0.20	0.25	0.30	+0.75 / −0.35
50	0.08	0.10	±0.12	±0.20	0.15	0.18	+0.60 / −0.20	0.25	0.30	+0.95 / −0.45
75	0.10	0.16	±0.15	±0.30	0.15	0.18	+0.70 / −0.30	0.25	0.35	+1.20 / −0.50
100	0.10	0.16	±0.20	±0.35	0.18	0.20	+0.85 / −0.35	0.25	0.35	+1.40 / −0.60
125	0.10	0.20	±0.25	±0.45	0.18	0.20	+1.00 / −0.40	0.25	0.40	+1.60 / −0.70
150	0.10	0.20	±0.30	±0.50	0.18	0.20	+1.10 / −0.50	0.25	0.40	+1.80 / −0.80
175	0.15	0.20	±0.30	±0.60	0.20	0.25	+1.25 / −0.55	0.25	0.40	+2.00 / −0.90
200	0.15	0.20	±0.35	±0.65	0.20	0.25	+1.40 / −0.60	0.25	0.40	+2.20 / −1.00
225	0.15	0.20	±0.40	±0.70	–	–	–	–	–	–
250	0.15	0.30	±0.40	±0.80	–	–	–	–	–	–
275	0.15	0.30	±0.45	±0.90	–	–	–	–	–	–
300	0.15	0.30	±0.50	±0.95	–	–	–	–	–	–
375	–	–	–	–	0.20	0.35	+2.40 / −1.00	0.25	0.50	+3.70 / −1.60
400	0.15	0.30	±0.65	±1.30	0.20	0.35	+2.50 / −1.10	0.25	0.50	+3.90 / −1.70
500	0.15	0.30	±0.80	±1.60	–	–	–	–	–	–
575	–	–	–	–	0.20	0.40	+3.50 / −1.50	0.25	0.70	+5.40 / −2.30
600	0.15	0.30	±0.95	±1.90	0.20	0.40	+3.65 / −1.55	0.25	0.70	+5.60 / −2.40
700	0.15	0.30	±1.10	±2.20	–	–	–	–	–	–
775	–	–	–	–	0.20	0.50	+4.60 / −2.00	0.25	0.80	+7.10 / −3.00
800	0.15	0.30	±1.25	±2.50	–	–	–	–	–	–
900	0.15	0.30	±1.40	±2.80	–	–	–	–	–	–
1000	0.15	0.30	±1.55	±3.10	–	–	–	–	–	–
1200	0.15	0.30	±1.85	±3.70	–	–	–	–	–	–

Reference and calibration grades have completely plain end faces and must be calibrated by the National Physical Laboratory.

Grade 1 bars are intended for use in inspection rooms and tool rooms.

Grade 2 bars are intended for use in measuring gauges, jigs, workpieces, etc.

Grades 1 and 2 have internally threaded ends, fig. 5.8, so that each bar can be used in combination with another by means of a freely fitting connecting screw. These threaded connections are assembled hand-tight only. In order to obtain specific lengths, gauge blocks can be wrung to the end faces of the length bar.

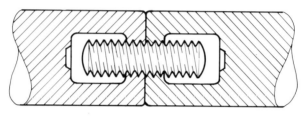

Fig. 5.8 Internally threaded ends of length bars

Length bars screwed together in combination and intended for use in the horizontal position should be supported at two symmetrically placed points a calculated distance apart, known as Airy points. If a bar is supported at two points, the sag caused by its own weight will affect the length between the end faces. In 1922, the then Astronomer Royal, Sir George Airy, established a formula for the distance separating two supports such that the upper and lower surfaces at the ends of the bar would lie in a horizontal plane and so give a true length between the end faces. This distance is 0.577 of the length of the bar, fig. 5.9. All bars 150 mm in length and over have the Airy points indicated by means of two symmetrically spaced pairs of lines engraved around the bars. For any combination over 150 mm, the Airy points are calculated and the engraved lines on individual bars are ignored.

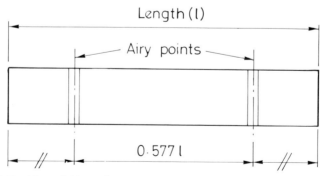

Fig. 5.9 Position of Airy points

Length-bar accessories The application of length bars can be extended by the use of a range of accessories. The specification for these is also contained in BS 5317:1976.

A typical set of length-bar accessories is shown in fig. 5.7 and is for use with the grade 1 and 2 bars having internally threaded ends. This set consists of

i) a base 25 mm thick, with the opposite faces finished by precision lapping to a high degree of surface finish, size, flatness, and parallelism. Length-bar combinations can be assembled to the base using a connecting screw in one of the threaded holes to give good stability when length bars are to be used vertically, fig. 5.10(a).

ii) a pair of large radiused jaws used in conjunction with a length-bar combination to form an internal or external caliper, fig. 5.10(b). One face is finished flat and the other is radiused by precision lapping to a width of 25 mm. The jaws have a plain bore and are assembled using a connecting screw and a knurled nut. One jaw can be used in conjunction with a length-bar combination and the base to assemble a height gauge.

iii) a pair of small plane-faced jaws used in conjunction with a length-bar combination to assemble a precise setting gauge, fig. 5.10(c). The two opposite wringing faces are precision-lapped to a high degree of surface finish, flatness, parallelism, and thickness. These jaws also have

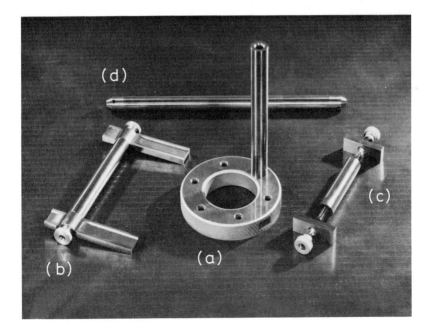

Fig. 5.10 Assembled length bars and accessories

a plain bore and are assembled using a connecting screw and a knurled nut.
iv) a pair of spherical end pieces, 25 mm in length, used in conjunction with a length-bar combination to assemble a precise setting rod or an internal measuring pin, fig. 5.10(d). The bore is threaded internally and is assembled using a connecting screw.
v) knurled nuts and connecting screws for assembly of length bars and accessories.

The accessories other than the nuts and connecting screws are made from high-grade steel, hardened and stabilised.

To obtain a specific size combination, gauge blocks can be inserted between the end of a length bar and the accessory. Two combinations of gauge blocks are required with any assembly, to fit between the length bar and the accessory at each side of the connecting screw. To provide these, special sets of gauge blocks are available for use with length bars. These sets contain two of each size of gauge block, each pair being matched for size to within 0.8 μm.

5.2 Straightness

Specific objective To state the principles of use and care of workshop standards for straightness.

The workshop standard against which the straightness of a line on a surface is compared is the straightedge. An error in straightness of a feature may be stated as the distance separating two parallel straight lines between which the surface of the feature, in that position, will just lie. Three types of straightedge are available: toolmaker's straightedges, cast-iron straightedges, and steel or granite straightedges of rectangular section.

Toolmaker's straightedges, covered by BS 852:1939(1959), are of short length up to 300 mm and are intended for very accurate work. They are made from high-quality steel, hardened and suitably stabilised, and have the working edge ground and lapped to a 'knife edge' as shown in the typical cross-section in fig. 5.11. Above 25 mm length, one end is finished at an angle. This type of straightedge is used by placing the knife edge on the work and viewing against a well illuminated background. If the work is

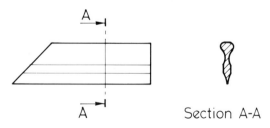

Fig. 5.11 Toolmaker's straightedge

perfectly straight at that position, then no white light should be visible at any point along the length of the straightedge. It is claimed that this type of test is sensitive to within 1 μm.

Cast-iron straightedges, of bow-shaped (fig. 5.12(a)) and I-section design (fig. 5.12(b)), are covered by BS 5204:part 1:1975. Two grades of accuracy are provided for each type – grade A and grade B – with grade A the more accurate. The straightedges are made from close-grained plain or alloy cast iron, sound and free from blowholes and porosity. The working surfaces of grade A are finished by scraping, and those of grade B by scraping or by smooth machining. The recommended lengths for the bow-shaped type are 300, 500, 1000, 2000, 4000, 6000, and 8000 mm and for the I-section type 300, 500, 1000, 2000, 3000, 4000, and 5000 mm.

Steel and granite straightedges of rectangular section are covered by BS 5204:part 2:1977. Two grades of accuracy are provided: grade A and grade B. Grade-A steel straightedges are made from high-quality steel with

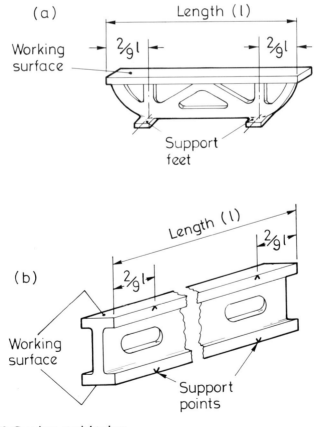

Fig. 5.12 Cast-iron straightedges

the working faces hardened. Grade-B steel straightedges may be supplied hardened or unhardened. Grade-A and grade-B straightedges may be made of close-grained granite of uniform texture, free from defects. The working faces are finished by grinding or lapping. The recommended lengths for rectangular-section straightedges are 300, 500, 1000, 1500, and 2000 mm.

When a straightedge is used on edge, it is likely to deflect under its own weight. The amount of deflection depends on the number and position of supports along the length of the straightedge. It has already been pointed out that length bars used in the horizontal position are supported at two points spaced apart a distance of 0.577 of the length of the bar. This brings the ends of the length bar horizontal and gives a true length between the end faces but it does not, however, give minimum deflection, which is the important condition for a straightedge. For minimum deflection, a straightedge must be supported at two points located two-ninths of its length from each end, fig. 5.12 – i.e. at supports symmetrically spaced five-ninths (0.555) of its length apart. For this reason, rectangular and I-section straightedges have arrows together with the word 'support' engraved on their side faces to indicate the points at which the straightedge should be supported for minimum deflection under its own weight.

If a rectangular or I-section straightedge is used on edge, it should not be placed directly on the surface being checked – it should be supported off the surface on two equal-size gauge blocks placed under the arrows marked 'support'. The straightness of the surface being checked is then established by determining the width of gap under the working face of the straightedge at various points along its length, using gauge blocks. Alternatively a dial indicator held in a surface gauge can be traversed along the surface being checked while in contact with the straightedge. Since the straightedge is straight, any deviation shown on the dial indicator will show the error of straightness of the surface.

Straightedges – especially the bow-shaped type – are used extensively to check the straightness of machine-tool slides and slideways. This is done by smearing a thin even layer of engineer's blue on the working surface of the straightedge, placing the straightedge on the surface to be checked, and sliding it slightly backwards and forwards a few times. Engineer's blue from the straightedge is transferred to the surface, giving an indication of straightness by the amount of blue present. Due to the width of the working face of the straightedge, an indication of flatness is also given.

5.3 Flatness

Specific objective To state the principles of use and care of workshop standards for flatness.

The workshop standard against which the flatness of a surface is compared is the surface plate or table. The error in flatness of a feature may be stated as the distance separating two parallel planes between which the surface of the feature will just lie. Thus flatness is concerned with the com-

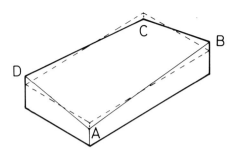

Fig. 5.13 Error in flatness of a surface

plete area of a surface, whereas straightness is concerned with a line at a position on a surface; for example, lines AB, BC, CD, and DA in fig. 5.13 may all be straight but the surface is not flat, it is twisted.

For high-precision work, such as precision tooling and gauge work, toolmaker's flats and high-precision surface plates are available and are covered by BS 869:1978. This standard recommends four sizes of toolmaker's flat – 63, 100, 160, and 200 mm diameter – made from high-quality steel hardened and stabilised or from close-grained granite of uniform texture, free from defects. Two sizes of high-precision flat are recommended – 250 and 400 mm diameter – made from plain or alloy cast iron or from granite. The working surface of flats and plates are finished by high-grade lapping and must be free from noticeable scratches and flat within 0.5 μm for flats up to 200 mm diameter, 0.8 μm for 250 mm diameter plates, and 1.0 μm for 400 mm diameter plates.

Surface plates and tables are covered by BS 817:1972, which specifies the requirements for rectangular and square surface plates ranging from 160 mm × 100 mm to 2500 mm × 1600 mm in three grades of accuracy – AA, A, and B – with grade AA the most accurate. The accuracy relates to the degree of flatness of the working surface.

The plates may be made of good-quality close-grained plain or alloy cast iron, sound and free from blowholes and porosity, and must have adequate ribbing on the underside to resist deflection. The working surface of grades AA and A is finished by scraping, while that of grade B may be finished by scraping or by machining. Alternatively, plates may be made of close-grained granite of uniform texture, free from defects and of sufficient thickness to resist deflection. The working surface must have a smooth finish.

The smaller sizes of plates may be used on a bench; the larger ones are usually mounted on a stand and are then known as surface tables.

The simplest method of checking the flatness of a surface is to compare it with a surface of known accuracy, i.e. a surface plate. This is done by smearing a thin even layer of engineer's blue on one surface, placing the surface to be checked on the surface plate, and moving it slightly from side to side a few times. Engineer's blue will be transferred from one surface to the other, the amount of blue present and its position giving an indication of the degree of flatness.

The main use of surface plates and tables is as a reference or datum surface upon which inspection and marking-out equipment are used.

5.4 Squareness

Specific objective To state the principles of use and care of workshop standards of squareness.

Two surfaces are square when they are at right angles to each other. Thus the determination of squareness is one of angular measurement. There is no absolute standard for angular measurement in the same way as there is for linear measurement, since the requirement is simply to divide a circle into a number of equal parts. The checking of right angles is a common requirement, and the workshop standard against which they are compared is the engineer's square, of which there are a number of types. BS 939:1977 specifies the requirements for engineer's try-squares (fig. 5.14(a)), cylindrical squares (fig. 5.14(b)), and block squares of solid (fig. 5.14(c)) or open form (fig. 5.14(d)).

Engineer's try-squares consist of a stock and a blade and are designated by a size which is the length from the tip of the blade to the inner working face of the stock. Recommended sizes are 50, 75, 100, 150, 200, 300, 450, 600, 800, and 1000 mm. Three grades of accuracy are specified – AA, A, and B – with grade AA the most accurate. Try-squares are made of good-quality steel, with the working surfaces of grades AA and A hardened and stabilised.

Grade AA try-squares have the inner and outer edges of the blade bevelled. All working surfaces of the blade and stock are lapped, finely ground, or polished to the accuracy specified for each grade.

Cylindrical squares, of circular section, are designated by their length. Recommended lengths are 75, 150, 220, 300, 450, 600, and 750 mm. One grade of accuracy is specified: grade AA. Cylindrical squares are made of high-quality steel, hardened and stabilised; close-grained plain or alloy cast iron, sound and free from blowholes and porosity; or close-grained granite of uniform texture, free from defects. Granite is particularly suitable for the larger sizes, as its mass is approximately half that of the equivalent size in steel or cast iron. In order to reduce weight, it is recommended that steel or cast-iron cylindrical squares 300 mm long and above are of hollow section. All external surfaces are finished by lapping or fine grinding.

Solid-form block squares are designated by their length and width and are available in sizes from 50 mm × 40 mm up to and including 1000 mm × 1000 mm. Two grades of accuracy are specified: AA and A. Solid-form block gauges are made of high-quality steel, cast iron, or granite – the same as cylindrical squares. Again, granite is recommended due to its lower mass. The front and back surfaces of each solid-form steel block square are recessed and fitted with a heat-insulating material, to avoid

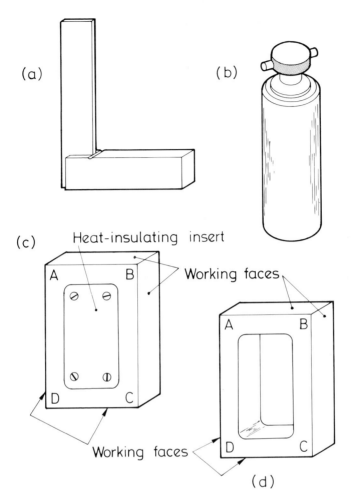

Fig. 5.14 Types of square

heat transfer and hence expansion when handled. The working faces of the solid-form steel block square are finished by lapping, and those made of cast iron or granite are finished by lapping or fine grinding.

Open-form block squares are designated by their length and width and are available in sizes from 150 mm × 100 mm up to and including 600 mm × 400 mm. Two grades of accuracy are specified – A and B – grade A being the more accurate. They are made of close-grained plain or alloy cast iron, sound and free from blowholes and porosity, and may be hardened or unhardened. All external surfaces are finished by lapping or fine grinding.

Use of squares
Grade AA engineer's try-squares have the inner and outer edges of the blade bevelled to produce a 'knife edge'. This increases the sensitivity of the square in use. If, however, the square is used with its blade slightly out of normal to the surface being checked, an incorrect result may be obtained. For this reason, try-squares with bevelled-edge blades are unsuitable for checking cylindrical surfaces. For this purpose a try-square with a square edge or a block square should be used, since the cylindrical surface itself will provide the necessary sensitivity by means of line contact.

Cylindrical squares are ideal for checking the squareness of try-squares and block squares and for work with flat faces, since line contact by the cylindrical surface gives greater sensitivity.

To check the squareness of two surfaces of a workpiece using a try-square, the stock is placed on one face and the edge of the blade is rested on the other. Any error is squareness can be seen by the amount of light between the surface and the underside of the blade. This type of check only tells whether or not the surfaces are square to each other, however, and it is difficult to judge the magnitude of any error.

When accurate results are required, the workpiece and square may be placed on a surface plate and the square slid gently into contact with the surface to be checked. The point of contact can then be viewed against a well illuminated background. If a tapering slit of light can be seen, the magnitude of the error present can be checked using gauge blocks at the top and bottom of the surface, fig. 5.15, the difference between the two gauge blocks being the total error of squareness.

5.5 Roundness

Specific objective   To describe with the aid of sketches a workshop test for roundness.

There are a number of workshop tests which are used to determine the roundness of a part; however, not all of these give a precisely true indication. A part is round when all points on its circumference are equidistant from its axis but, as a result of different methods and of machine tools used in the production of cylindrical parts, errors in roundness can occur.

The simplest check for roundness is to measure directly at a number of diametrically opposite points around the circumference of a part, using a measuring instrument such as a micrometer or vernier caliper. Any difference in reading will give an indication of the out-of-roundness of the part. However it is possible, when using this method, for errors in roundness to go undetected. For example, an incorrect set-up on a centreless grinding machine can produce a tri-lobed shape such as is shown exaggerated in fig. 5.16. Measuring at diametrically opposite points will give identical readings, but the part is not round.

Alternatively, a part which contains centres at each end may be checked for roundness by mounting between bench centres and rotating the part

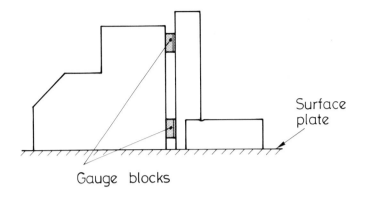

Fig. 5.15 Checking squareness with square and gauge blocks

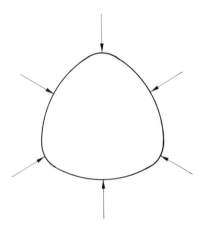

Fig. 5.16 Tri-lobed shape

Fig. 5.17 Checking work between centres

under a dial indicator, fig. 5.17. An error reading on the dial indicator would show the part to be not round. However this method can also be misleading, since the centres in the part themselves may not be round or may not be the central axis of the part and it may be these errors which are represented on the dial indicator and not the error in roundness of the part.

A part with a plain bore may be loaded on to a mandrel before being placed between the bench centres. In this case, since a mandrel is accurately ground between true centres, it can be assumed that the centres are on the true central axis. A constant reading on the dial indicator would show the part to be round. It is also an indication that both the bore and the outside diameter lie on the same axis – a condition known as concentricity. An error reading on the dial indicator could therefore be an error in concentricity and the part be perfectly round.

The ideal workshop test which overcomes the problems already outlined is to rotate the part under a dial indicator with the part supported in a vee block. Because the points of support in the vee block are not diametrically opposite the plunger of the dial indicator, errors in roundness will be identified. For example the tri-lobed condition undetected by direct measurement would be detected by this method, as shown in fig. 5.18.

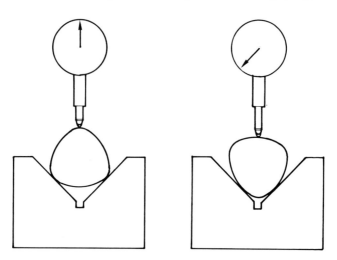

Fig. 5.18 Checking work in a vee block

5.6 The sine bar

Specific objective To use the sine bar for measurement.

The measurement of angles can be simply carried out by using a protractor, but the degree of accuracy obtainable is, at best, 5 minutes using a vernier instrument. Greater accuracy can be obtained by using a sine bar in conjunction with gauge blocks.

A number of different designs of sine bar are available and are covered by BS 3064:1978, which also specifies three lengths – 100, 200, and 300 mm – these being the distances between the roller axes. The more common type of sine bar is shown in fig. 5.19, the body and rollers being made of high-quality steel, hardened and stabilised, with all surfaces finished by lapping or fine grinding to the tolerances specified in the British Standard.

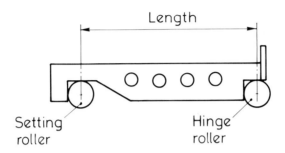

Fig. 5.19 Sine bar

The main requirements are

a) the mean diameters of the rollers shall be equal to each other within 0.0025 mm;
b) the upper working surface shall be parallel to the plane tangential to the lower surface of the rollers within 0.002 mm;
c) the distance between the roller axes shall be accurate to within

    0.0025 mm for 100 mm bars,
    0.005 mm for 200 mm bars,
    0.008 mm for 300 mm bars.

In use, gauge blocks are placed under the setting roller with the hinge roller resting on a datum surface, e.g. a surface plate. The angle of inclination is then calculated from the length of the sine bar (L) and the height of the gauge-block combination (h), fig. 5.20. Since the roller diameters are the same, we can consider triangle ABC, where AB = L and BC = h; thus

$$\sin \theta = \frac{BC}{AB} = \frac{h}{L}$$

If a known angle of inclination is required then the necessary gauge-block combination can be calculated from $h = L \sin \theta$. However, it is more usual to need to measure the angle of inclination, and in this case the object to be measured is placed on the upper working surface of the sine bar and a gauge-block combination is made up by trial and error. This is

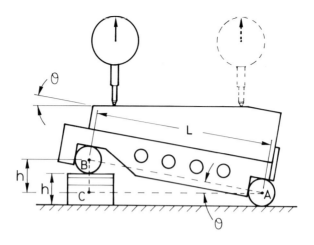

Fig. 5.20 Sine-bar set-up

done in conjunction with a dial indicator mounted in a surface gauge, the correct height of gauge blocks being established when the dial indicator gives the same reading at each end of the object being measured. The angle is then calculated from $\sin \theta = h/L$.

The degree of accuracy obtainable using a sine bar depends, among other factors, on the size of angle being checked. For example, using a 200 mm sine bar, the following gauge-block heights are required:

for an angle of 10°   = 34.73 mm
      for 10° 1′ = 34.79 mm
      for 60°   = 173.20 mm
      for 60° 1′ = 173.23 mm

Thus for a 1′ difference at 10° a variation in gauge-block height of 0.06 mm is required, while the same 1′ difference at 60° requires a 0.03 mm variation.

This means, in theory at least, that if the smallest increment of gauge block available is 0.001 mm then, using a 200 mm sine bar, this represents an angular difference of 1 second at small angles and an angular difference of 2 seconds at around 60°. In practice, however, these accuracies would not be possible under workshop conditions, due to inaccuracies in gauge blocks, the sine bar, setting-up, and using the dial indicator; but it can be appreciated that this simple piece of workshop equipment is extremely accurate and capable of measuring angles within 1 minute.

It can be seen from the above that the larger angles cannot be measured to the same degree of accuracy as smaller angles. Because of this and the fact that the larger angles require a large gauge-block combination which makes the complete set-up unstable, it is recommended that for larger angles the complement of the angle be measured, i.e. 90° minus the required angle.

Other equipment based on the same principle is available – sine tables with a larger working surface and inclinable about a single axis for checking larger work, compound sine tables inclinable about two axes for checking compound angles, and sine centres which have centres in each end and are the most convenient for checking taper work.

5.7 Calibrated steel balls and rollers

Specific objective To use calibrated steel balls and rollers for measurement.

In conjunction with gauge blocks and measuring equipment such as micrometers, calibrated steel balls and rollers can be used for the accurate measurement of a wide range of features in the workshop.

Calibrated steel balls are available in sets, a typical example of which would contain sizes from 1 mm to 25 mm diameter in increments of 1 mm, fig. 5.21. Each size in the set contains three balls matched for uniformity of size.

Calibrated steel rollers are also available in sets, a typical example of which would contain sizes from 5 mm to 10 mm diameter in increments of

Fig. 5.21 Set of calibrated steel balls

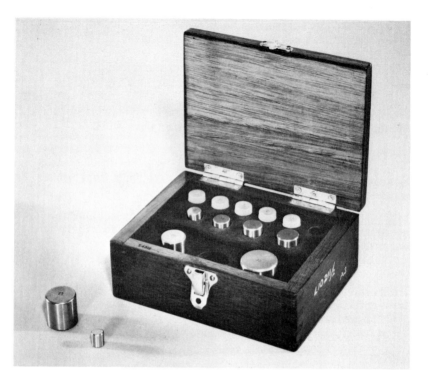

Fig. 5.22 Set of calibrated steel rollers

1 mm or 12 mm to 20 mm diameter in increments of 2 mm, fig. 5.22. Each size in the set contains two rollers matched for uniformity of size. Sets of balls and rollers are calibrated to show accuracy of size.

Measurement of plain bores
Balls and rollers can be used in conjunction with other measuring equipment to accurately check the diameter of plain bores. The simplest method is to use two rollers together with a gauge-block combination built-up so that it just slides between the two rollers placed diametrically opposite each other against the sides of the bore, fig. 5.23(a). Normally two rollers of the same size are used, the diameter of the bore being equal to twice the diameter of the roller plus the gauge-block combination.

Smaller bores where there is insufficient space to insert rollers and gauge blocks can be measured using two balls, fig. 5.23(b). These are chosen such that the centre of the upper ball contacts the side of the bore at the highest position possible. In this position, the top of the upper ball will be above the level of the work. The arrangement is placed on a surface plate and the height h is measured. This may be done by placing a gauge-block combination alongside and using a dial indicator mounted on a surface

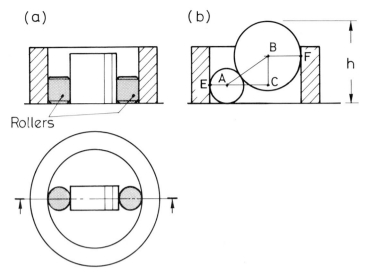

Fig. 5.23 Measurement of plain bore

gauge, checking from the top of the upper ball to the gauge blocks. Alternatively, a depth micrometer raised off the surface of the work on gauge blocks can be used. Having established height h, the diameter of the bore can be calculated from triangle ABC in fig. 5.23(b).

Let diameter of large ball = D

diameter of small ball = d

Diameter of bore = EA + AC + BF

but EA = $\dfrac{d}{2}$ and BF = $\dfrac{D}{2}$

also AB = $\left(\dfrac{D + d}{2}\right)$ and BC = $h - \left(\dfrac{D + d}{2}\right)$

Using Pythagoras,

$$\begin{aligned}
AC^2 &= AB^2 - BC^2 \\
&= \left(\dfrac{D + d}{2}\right)^2 - \left[h - \left(\dfrac{D + d}{2}\right)\right]^2 \\
&= \left(\dfrac{D + d}{2}\right)^2 - \left[h^2 - h(D + d) + \left(\dfrac{D + d}{2}\right)^2\right] \\
&= \left(\dfrac{D + d}{2}\right)^2 - h^2 + h(D + d) - \left(\dfrac{D + d}{2}\right)^2 \\
&= h(D + d) - h^2
\end{aligned}$$

∴ AC = $\sqrt{h(D+d) - h^2}$

But diameter of bore = EA + AC + BF

$$= \frac{d}{2} + \sqrt{h(D+d) - h^2} + \frac{D}{2}$$

$$= \left(\frac{D+d}{2}\right) + \sqrt{h(D+d) - h^2}$$

The same result can be obtained using four balls. Three balls of the same diameter are placed in the bore with a larger-diameter ball placed on top, fig. 5.24. The small balls contact the side of the bore while the large one contacts all three. The height h is measured as before, using a dial indicator or depth micrometer. The diameter of the bore can be calculated from triangle ABC in fig. 5.24.

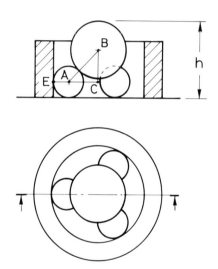

Fig. 5.24 Measurement of plain bore

Let diameter of large ball = D

  mean diameter of small balls = d

Diameter of bore = 2 (EA + AC)

but EA = $\frac{d}{2}$

also AB = $\left(\frac{D+d}{2}\right)$ and BC = $h - \left(\frac{D+d}{2}\right)$

Using Pythagoras,

$$AC^2 = AB^2 - BC^2$$

$$= \left(\frac{D+d}{2}\right)^2 - \left[h - \left(\frac{D+d}{2}\right)\right]^2$$

$$= \left(\frac{D+d}{2}\right)^2 - \left[h^2 - h(D+d) + \left(\frac{D+d}{2}\right)^2\right]$$

$$= \left(\frac{D+d}{2}\right)^2 - h^2 + h(D+d) - \left(\frac{D+d}{2}\right)^2$$

$$= h(D+d) - h^2$$

$$\therefore AC = \sqrt{h(D+d) - h^2}$$

But diameter of bore $= 2(EA + AC)$

$$= 2\left(\frac{d}{2} + \sqrt{h(D+d) - h^2}\right)$$

$$= d + 2\sqrt{h(D+d) - h^2}$$

Measurement of internal-taper features
Balls and rollers can be used in conjunction with other measuring equipment to accurately check the angle of internal-taper features such as taper ring gauges, morse tapers, and dovetail slideways.

For larger taper bores, two balls of the same size are placed diametrically opposite each other at the bottom of the bore, resting on a surface plate. The dimension between them, M_1, is established using a gauge-block combination which will just slide between the two balls. The two balls are then raised near to the top of the taper bore while resting on two gauge blocks of the same height h such that the centre of the balls is just below the

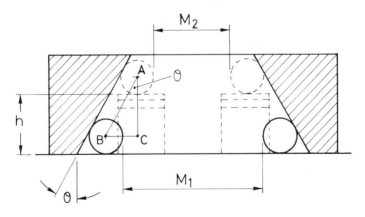

Fig. 5.25 Measurement of large taper bore

top surface, fig. 5.25. The dimension between the balls, M_2, is again established using a gauge-block combination. The angle of taper is calculated from triangle ABC in fig. 5.25.

$$\tan \theta = \frac{BC}{AC} \quad \text{where} \quad \theta = \text{half the included angle}$$

Let diameter of balls = D

Since the balls are the same diameter, AC = h

$$BC = \frac{(M_1 + D) - (M_2 + D)}{2}$$

$$= \frac{M_1 + D - M_2 - D}{2}$$

$$= \frac{M_1 - M_2}{2}$$

$$\therefore \tan \theta = \frac{BC}{AC} = \frac{M_1 - M_2}{2h}$$

Since θ equals half the angle of taper, this must be doubled to find the included angle. The same principle can be used to find the diameter at the mouth of the taper bore.

For smaller-diameter taper bores, only two balls need be used, fig. 5.26. In this case the small ball is placed in the bore and the dimension from the top face h_1 is found using a depth micrometer. A larger diameter ball is placed in the bore and the dimension h_2 is found using a depth micrometer. The angle of taper is calculated from triangle ABC in fig. 5.26.

$$\sin \theta = \frac{BC}{AB}$$

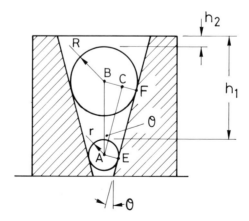

Fig. 5.26 Measurement of small taper bore

Let diameter of large ball = D and radius = R
     diameter of small ball = d and radius = r

BC = $R - r$

AB = $(h_1 + r) - (h_2 + R)$

$\therefore\ \sin\theta = \dfrac{R - r}{(h_1 + r) - (h_2 + R)} = \dfrac{R - r}{(h_1 - h_2) - (R - r)}$

Since θ equals half the angle of taper, this must be doubled to find the included angle. The same principle can be used to find the diameter at the mouth of the taper bore.

The angle of internal dovetail slideways can be checked using rollers in conjunction with gauge blocks or vernier calipers. Two small-diameter rollers of the same size are placed in the bottom of the angle and the distance between them, M_1, is measured using a gauge-block combination, fig. 5.27. If there is insufficient space for gauge blocks, this distance can be measured using the external jaw of a vernier caliper. Two larger-diameter rollers, again of the same size, are inserted in the angle and the distance between them, M_2, is measured using a gauge-block combination. The angle of taper can be calculated from triangle ABC in fig. 5.27.

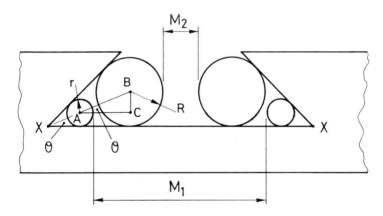

Fig. 5.27 Measurement of internal dovetail

$\tan\theta = \dfrac{BC}{AC}$

Let diameter of large roller = D and radius = R
     diameter of small roller = d and radius = r
then BC = $R - r$

and $AC = \dfrac{(M_1 + d) - (M_2 + D)}{2}$

$\therefore \tan \theta = \dfrac{BC}{AC} = \dfrac{R - r}{[(M_1 + d) - (M_2 + D)]/2}$

$= \dfrac{2(R - r)}{(M_1 + d) - (M_2 + D)}$

$= \dfrac{D - d}{(M_1 + d) - (M_2 + D)} = \dfrac{D - d}{(M_1 - M_2) - (D - d)}$

Since θ equals half the angle of taper, this must be doubled to find the included angle. The same principle can be used to find the dimension between points XX.

Measurement of external-taper features

Rollers can be used in conjunction with other measuring equipment to accurately check the angle of external-taper features such as taper plug gauges, morse tapers, and dovetail slides.

Taper plug gauges can be placed on a surface plate with two rollers of the same size placed diametrically opposite each other in contact with the gauge, fig. 5.28. The dimension M_1 is measured using a micrometer. The same two rollers are then raised on two equal gauge-block combinations of height h such that their centre is just below the top of the gauge. The dimension M_2 is measured using a micrometer. The angle of taper is calculated from triangle ABC in fig. 5.28.

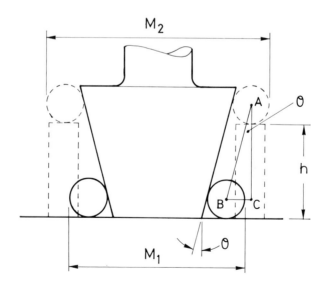

Fig. 5.28 Measurement of external taper plug

$$\tan \theta = \frac{BC}{AC}$$

Let diameter of rollers = D

Since the rollers are the same diameter, $AC = h$

$$BC = \frac{(M_2 - D) - (M_1 - D)}{2}$$

$$= \frac{M_2 - D - M_1 + D}{2}$$

$$= \frac{M_2 - M_1}{2}$$

$$\therefore \tan \theta = \frac{BC}{AC} = \frac{M_2 - M_1}{2h}$$

Since θ equals half the angle of taper, this must be doubled to find the included angle. The same principle can be used to find the diameter of the small end of the gauge.

The angle of external dovetail slides can be checked using rollers in conjunction with a micrometer, fig. 5.29. Two small-diameter rollers of the same size are placed in the bottom of the angle and the distance between them, M_1, is measured using a micrometer. Two larger diameter rollers, again of the same size, are then inserted in the angle and the distance between them, M_2, is measured using a micrometer. The angle of taper can be calculated from triangle ABC in fig. 5.29.

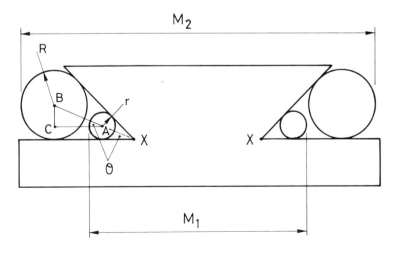

Fig. 5.29 Measurement of external dovetail

$$\tan \theta = \frac{BC}{AC}$$

Let diameter of large roller $= D$ and radius $= R$

diameter of small roller $= d$ and radius $= r$

then $BC = R - r$

and $AC = \dfrac{(M_2 - D) - (M_1 - d)}{2}$

$\therefore \tan \theta = \dfrac{BC}{AC} = \dfrac{R - r}{[(M_2 - D) - (M_1 - d)]/2}$

$= \dfrac{2(R - r)}{(M_2 - D) - (M_1 - d)}$

$= \dfrac{D - d}{(M_2 - D) - (M_1 - d)} = \dfrac{D - d}{(M_2 - M_1) - (D - d)}$

Since θ equals half the angle of taper, this must be doubled to find the included angle. The same principle can be used to find the dimension between points XX.

5.8 Comparative measurement

Specific objective To describe the principles of comparative measurement.

Comparative measurement is where the size of a workpiece is established by comparing it with that of a known standard, such as gauge blocks, length bars, or setting gauges. The instrument used to establish any difference in size between the standard and the work is first set up using the required size of standard. The instrument is then used to compare the work with this standard, any small difference in size being magnified and shown as a visual display.

The simplest form of comparative measurement in the workshop is carried out using a variety of equipment incorporating a dial indicator which, depending on the application and the accuracy required, can be readily set by micrometers and gauge blocks.

Comparative measurement is a convenient means of measuring and can usually be done with consistency, more quickly than direct measurement, and with greater accuracy. For example, a micrometer can be read to 0.01 mm or a micrometer with a vernier scale to 0.002 mm, while a vernier caliper can be read to 0.02 mm. Dial indicators can have an accuracy as high as 0.001 mm easily readable on a large dial.

In some cases, comparative measurement can be simply carried out where direct methods are difficult and lengthy.

5.9 Sources of error

Specific objective To describe how sources of error in the direct use of instruments are overcome by comparative methods of measurement.

The main source of error in the use of direct-measuring instruments arises from 'feel' – or rather the lack of it – i.e. the ability to judge the force being applied during measuring. This 'feel' can be developed only from experience of using the particular instrument, e.g. vernier calipers and internal micrometers, being more difficult with some instruments than with others.

This problem is overcome in comparative measurement, in the case of those instruments incorporating a dial indicator, by springs within the mechanism which apply a constant force at the stylus or contact point. This gives the same consistent 'feel' during setting up and during use.

Another source of error is human error where mistakes in reading a direct-measuring instrument can arise. This is usually the result of carelessness, although some instruments – particularly the vernier caliper – may be difficult to read. Nevertheless, errors can occur.

Human error is overcome in comparative measurement by providing a large dial so that in operation it is a simple matter to relate the position of a pointer and the zero mark. This gives an assured reliability of result.

5.10 Methods of comparative measurement

Specific objective To describe methods of comparative measurement for length and diameter.

The simplest set-up for the comparative measurement of length and outside diameter is a dial indicator mounted in a surface gauge set on a surface plate. A gauge-block combination of the required size is placed on the surface plate, and the dial indicator is lowered until the stylus or contact point touches and gives a reading. It is not necessary to line up the pointer exactly with zero, as the dial on these instruments can be rotated to bring zero in line with the pointer. The gauge blocks are then withdrawn and replaced by the workpiece to be measured. Any difference in reading from that obtained with the gauge blocks is the amount of error and may be read plus or minus.

With such a set-up, it is essential to ensure maximum rigidity in mounting the dial indicator. Excessive overhang must be avoided.

Where greater flexibility is required without the need to use a surface plate and the set-up just described, comparator stands complete with a working surface are available for use with dial indicators. A typical example is shown in fig. 5.30. These are of rigid design and can be used anywhere in the workshop.

With plunger-type indicators it is important to have the plunger vertical in both planes.

With stylus-type indicators fitted with a spherical point it is important that the stylus be kept horizontally in line with the indicator pivot – any

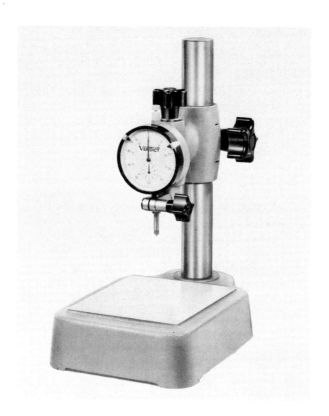

Fig. 5.30 Comparator stand and dial indicator

angle will affect the value of the graduations on the dial. As shown in fig. 5.31, the distance from the point of contact to the indicator pivot is x. When the stylus is tilted at an angle as shown, the distance from the point of contact to the indicator pivot is reduced to y, where $y = x \times$ the cosine of the angle. For example, for an indicator with a scale range of 0.8 mm

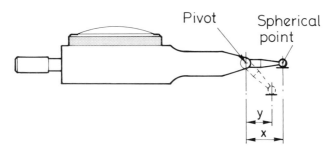

Fig. 5.31 Lever-type dial indicator with spherical-point stylus

and a stylus 14 mm long tilted at 30°, the effective length of the stylus is reduced to 14 cos 30° = 12.12 mm. If the 14 mm long stylus gives a range of 0.8 mm per revolution of the pointer, a stylus length of 12.12 mm will only give a range of 0.69 mm, giving an error of 0.11 mm per revolution of the pointer. Any error will increase with an increase in angle and in the length of stylus.

This problem is overcome by the use of a pear-shaped point, fig. 5.32. The effective length of this remains constant even when tilted within a maximum of 36°, and the indicator will still read correctly. This is shown diagramatically in fig. 5.33.

Fig. 5.32 Lever-type dial indicator with pear-shaped-point stylus

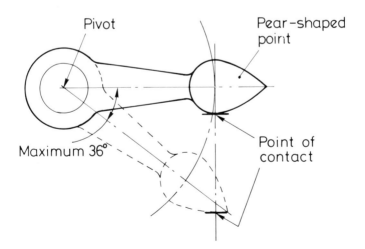

Fig. 5.33 Contact position of pear-shaped-point stylus

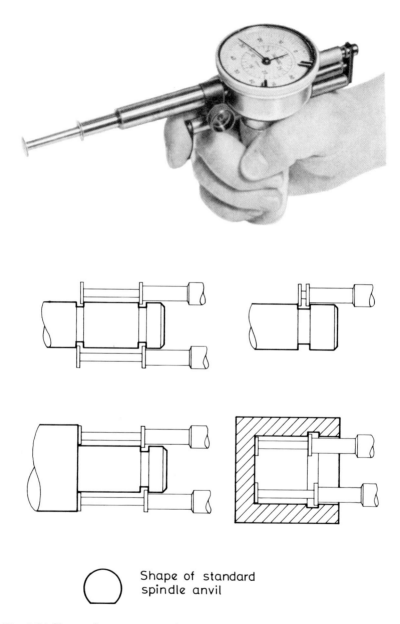

Fig. 5.34 Slot- and groove-measuring gauge

Comparative measurement of length and external diameter can be carried out using simple hand-held equipment. The measuring gauge shown in fig. 5.34 is used to measure easily and quickly the length between faces of slots and grooves, both external and internal – an operation which in some cases could be difficult and lengthy using measuring equipment such as gauge blocks, a micrometer, or a vernier caliper. The model shown has a measuring range up to 60 mm. External dial calipers as shown in fig. 5.35 are available in a range of leg lengths and measuring ranges and can be easily used to measure diameter and length – particularly internal wall thickness, which may not be easy using a micrometer or vernier caliper. These may be set to the required size using gauge blocks, the legs being operated by means of the button on the side of the instrument.

Fig. 5.35 External dial calipers

Internal measuring of bores and recesses often presents problems when using equipment such as internal micrometers and vernier calipers. It is difficult to obtain a consistent 'feel' using internal micrometers, especially at any great distance along the bore. Vernier calipers are limited to measuring the diameter of the mouth of the bore only, due to the short length of the jaws. Neither instrument is suitable for measuring the diameter of an internal recess. The internal dial calipers shown in fig. 5.36 overcome the problem of 'feel' and are capable of measuring the diameter of an internal recess. These instruments are available in a range of leg lengths and measuring ranges. They can be easily set to read a required size using an external micrometer, a setting piece, or gauge blocks used with suitable accessories. In operation, the legs are closed by means of a button on the side and are inserted in the bore; the button is then released and the reading is taken from the indicator.

Fig. 5.36 Internal dial calipers

High-precision internal measuring of bores can be carried out using an internal comparator known as a bore gauge. Bore gauges are available in a range of sets capable of measuring bores from 0.47 mm to 41 mm diameter to an accuracy of 0.001 mm. Unit heads, each covering a small range of measurement, are interchangeable with a dial indicator and extension rods to suit a variety of diameters and of bore lengths up to about 750 mm long. The desired size may be set using a micrometer, gauge blocks with jaw accessories, or, for greater accuracy, by using a precision setting ring. A typical bore gauge is shown in fig. 5.37. Precision bore gauges employing a different style of gauge head are available covering a range of diameters up to 350 mm.

In use the bore gauge, once set, is inserted in the bore being measured and is rocked backwards and forwards until the maximum reading is shown on the dial indicator. This eliminates the difficulty of obtaining the correct 'feel' often encountered using direct-measuring instruments.

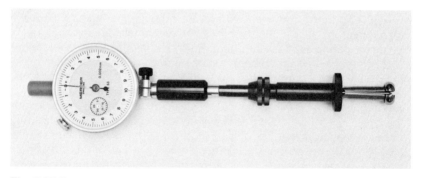

Fig. 5.37 Bore gauge

Exercises on chapter 5

1 What is meant by 'comparative measurement'?
2 State the five grades of gauge block and indicate where each would be used.
3 Name three materials from which straightedges are made.
4 Where are the UK physical standards maintained?
5 Describe an ideal workshop test for roundness.
6 What is the function of the British Calibration Service?
7 If length bars are used horizontally, they should be supported at two symmetrically placed points. What are these points called and what is their position?
8 Give a reason why standards are necessary in industry.
9 Name three types of square used in industry.
10 A 200 mm sine bar is used to check an angle of 30° 12′. Calculate the size of gauge blocks necessary. [*Answer*: 100.604 mm]
11 At what temperature are gauge blocks calibrated and intended for use?
12 The angle of a taper plug gauge is to be checked using two 10 mm diameter rollers. The measurement across the two rollers at the small end of the gauge is 45.38 mm. The two rollers are raised to the large end on 90 mm gauge blocks and the measurement across the rollers is then found to be 83.64 mm. Calculate the angle of the taper plug gauge. [*Answer*: 24°]

6 Single-point metal cutting

General objective The student understands the basic principles of the metal-cutting process using single-point tools.

A single-point cutting tool is defined as a tool which terminates in a single cutting point. The functional part of the tool employed in cutting consists of the major cutting edge, which removes most of the material and takes the greatest share of the cutting load, and the minor cutting edge which is mainly responsible for producing the finished workpiece surface, fig. 6.1.

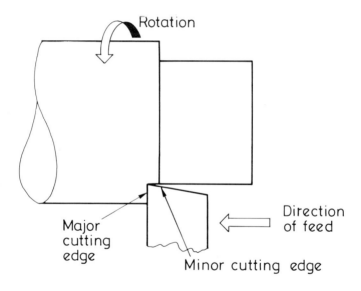

Fig. 6.1 Single-point cutting

Typical single-point cutting tools are those used for turning and shaping operations, some typical shapes of which are shown in fig. 6.2. Where appropriate, the same shapes are used both for turning and shaping, the only difference being the size of shank, which is thicker in section for shaping operations due to the greater shock loads encountered.

The way in which metal cutting takes place is a function of the type of material being machined and the cutting conditions, e.g. cutting speed, feed rate, depth of cut, and tool geometry.

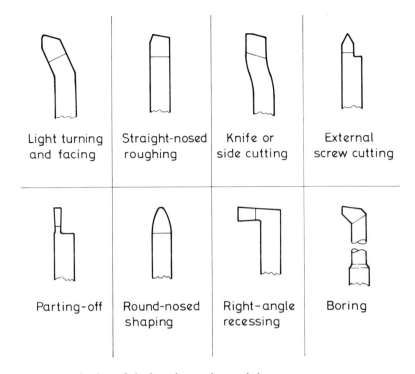

Fig. 6.2 A selection of single-point cutting-tool shapes

The way in which a chip is formed in metal cutting is by deformation of the metal ahead of the cutting edge by a process of shear. The sheared metal then starts to flow along the face of the tool, i.e. the metal compresses and then shears. Three basic types of chip formation are generally recognised, irrespective of the type of machining operation:

i) discontinuous chip,
ii) continuous chip,
iii) continuous chip with built-up edge.

Discontinuous chip (fig. 6.3)   This type of chip consists of individual segments produced by actual fracture of the metal ahead of the cutting edge. It is most often found in the machining of brittle materials, when it results in a fair surface finish and reasonable tool life. It can also occur in the machining of ductile materials at very low cutting speeds, using small rake angles, but it then results in a poor surface finish and excessive tool wear.

Continuous chip (fig. 6.4)   This type of chip is formed by continuous deformation of the metal ahead of the cutting edge without fracture,

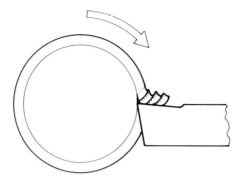

Fig. 6.3 Discontinuous chip

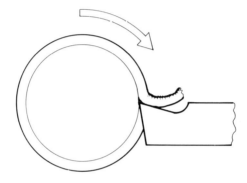

Fig. 6.4 Continuous chip

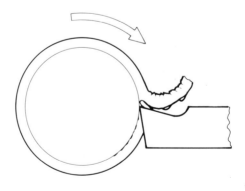

Fig. 6.5 Continuous chip with built-up edge

followed by smooth flow of the chip along the tool face due to low friction between the chip and the tool. It is the most desirable type and is found in the machining of ductile materials at high cutting speeds with larger rake angles and results in good surface finish and tool life. Long continuous chips can, however, be dangerous to the machine operator and can tangle in the chuck or workpiece. Some form of chip-breaker may be necessary to break the chips into short lengths for ease of disposal.

Continuous chip with built-up edge (fig. 6.5)   This type of chip is similar to the continuous chip except that a built-up edge is present on the tool face. The metal shears without fracture, but the resistance to sliding along the tool face produces additional shearing of the chip. A part of the underside of the chip shears away and remains on the tool face as a built-up edge. The size of the built-up edge continually changes during cutting as more metal shears from the flowing chip until a point is reached where the built-up edge eventually breaks down, part of it going off with the chip and part of it being deposited on the workpiece surface. This building up and breaking down occurs very rapidly during machining and results in a poor surface finish.

The built-up edge is common in metal-cutting operations and is found in the machining of ductile materials with high-speed-steel tools at ordinary cutting speeds. Its effects can be minimised by increasing the rake angle of the tool, increasing the cutting speed, and reducing friction between chip and tool by the use of a cutting fluid to provide lubrication and by having a high surface finish on the tool face.

6.1 Cutting force

Specific objective   To state the basic effects of variations in speed, feed, depth of cut, and tool geometry on cutting force.

There are two ways in which the major cutting edge of a single-point cutting tool can be presented to the workpiece:

i) where the cutting edge is perpendicular to the direction of feed, known as orthogonal cutting, fig. 6.6;
ii) where the cutting edge is at an angle to the direction of feed, known as oblique cutting, fig. 6.7.

With orthogonal cutting, the force acting at a point on the cutting edge can be resolved into two components as shown in fig. 6.6, which shows a turning operation using a knife tool. The force of the chip on the cutting tool acts vertically down, tangential to the work, and is known as the cutting force. The second force, acting longitudinally along the axis of the work as a reaction to the direction of feed, is known as the feed force.

With oblique cutting, the force acting can be resolved as shown in fig. 6.7. The cutting force acts vertically in the same way as in orthogonal cutting. Due to the cutting edge being at an angle, two component forces are resolved: one reacting against the direction of feed as before and a radial force tending to push the tool radially away from the workpiece.

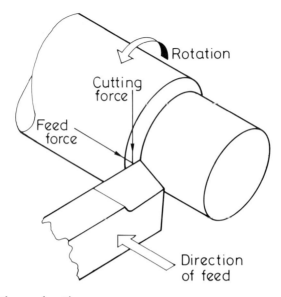

Fig. 6.6 Orthogonal cutting

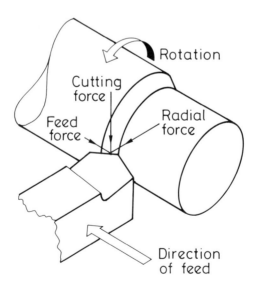

Fig. 6.7 Oblique cutting

The major force acting during metal cutting is the cutting force, and the effects on this force of variations in cutting speed, feed, depth of cut, and tool geometry can be established from practical cutting tests using a tool-force dynamometer, which is covered in *Manufacturing technology for level-3 technicians*. Cutting force is independent of cutting speed within the practical range of cutting speeds normally used.

Any increase in feed rate increases the cutting force, as does any increase in depth of cut. The effect of a change in depth of cut on the cutting force, however, is greater than the effect of a change in feed rate.

The aspect of tool geometry having the greatest influence on cutting force is the rake angle, i.e. the angle between the tool face and a line at right angles to the surface of the material being cut (see section 6.5). Any increase in rake angle reduces the cutting force. Thus for negative rake angles the cutting forces are high, but the cutting forces reduce as the rake angle is increased in a positive direction. There is naturally a limit to any positive rake angle, since the tool point becomes progressively weaker as the angle is increased, becoming more prone to breakdown.

6.2 Tool life

Specific objective To state the basic effects of variations in speed, feed, depth of cut, and tool geometry on tool life.

Tool life may be defined as the time between two successive regrinds during which the tool was actually cutting metal. It can be judged by the amount of wear which takes place on the flank face of the tool. When a predetermined size of wear land (i.e. the depth of the worn portion on the flank face) is reached, the cutting edge is resharpened or is replaced by a new tool or tip.

Tool life is an important factor in metal cutting, and conditions have to be chosen which will give an economic tool life. Those conditions which give a short tool life will not be economic, since tool-grinding and tool-replacement costs will be high. On the other hand, conditions which give a long tool life will also not be economic, since metal-removal rates will be low.

The variable that has the greatest influence on tool life is cutting speed. The relationship between tool life and cutting speed is represented by the formula

$$VT^n = C$$

where V = velocity or cutting speed in m/min

T = tool life in minutes

C = a constant for given tool and workpiece materials and cutting conditions

n = an exponent for given tool and workpiece materials and cutting conditions

Calculations using this formula are dealt with in *Manufacturing technology for level-3 technicians*.

It can be shown that tool life decreases as cutting speed is increased. This is due to the higher heat generation with increasing cutting speed. This increased heat generation has less effect on those cutting-tool materials having a high red-hardness value (i.e. the ability of a cutting-tool material to retain its hardness at high cutting temperatures), thus ceramics and cemented carbides are less affected than high-speed steel.

The cutting speed has a much greater effect on the amount of heat generated than does feed rate or depth of cut. Because of this, greater metal-removal efficiency is obtained by increasing the feed rate and depth of cut. However, in order to maintain a constant tool life and take advantage of the greater metal-removal efficiency, any increase in feed rate and depth of cut must be accompanied by a decrease in cutting speed. The decrease in cutting speed is not proportional to an increase in feed rate or depth of cut, but approximate relationships are shown in figs 6.8 and 6.9. As can be seen, if the feed rate is doubled (i.e. 100% increase), the cutting speed should be reduced by 30% to give the same tool life. Also, if the depth of cut is doubled (i.e. 100% increase), the cutting speed should be reduced by 20% to give the same tool life.

As a general rule, it can be stated that the combination of a high feed rate and a large depth of cut with a low cutting speed will allow a large volume of metal to be removed during a given tool life.

As previously stated, the heat generated during cutting has a marked effect on tool life. If the rake angle is increased, less heat is generated.

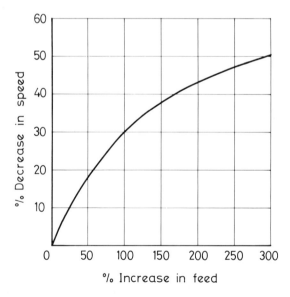

Fig. 6.8 Approximate relationship between feed and speed for given tool life

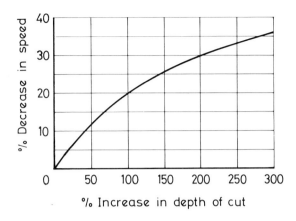

Fig. 6.9 Approximate relationship between depth of cut and speed for given tool life

However, the increased rake angle weakens the tool and reduces the rate of heat dissipation and so shortens the life of the tool.

If the rake angle is large, the heat is concentrated in a small area of the tool, fig. 6.10(a). If this area is too small to dissipate the heat, the temperature may rise to an extent which will affect the hardness of the tool material. If the rake angle is decreased, the heat-dissipation area is increased, fig. 6.10(b), and the temperature is kept low enough to prevent softening of the tool material, leading to an improvement in tool life. This is especially important with high-speed steel. The decrease in rake angle also increases the strength of the tool tip, as discussed in section 6.5.

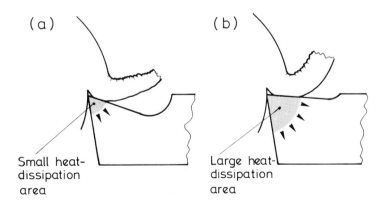

Fig. 6.10 Effect of rake angle on heat dissipation

6.3 Power consumption

Specific objective   To state the basic effects of variations in speed, feed, depth of cut, and tool geometry on power consumption.

In machining, the total power available is needed both to overcome the force at the cutting tool during metal removal and also to overcome frictional losses in the machine transmission. Thus only part of the power supplied by the machine motor is available for metal removal.

The difference between the total power available and the amount used in metal removal is the power required to overcome frictional losses in the machine. From this, the efficiency of the machine can be calculated.

Power is the product of force and velocity. In metal cutting, the velocity is the cutting speed. The majority of the power used in actual metal removal is required to overcome the cutting force. Although the feed force may be quite high, the feed velocity is so low that the power required to feed the tool can be neglected. The radial force does not affect power since there is no velocity in a radial direction.

As previously stated, the cutting force is affected by variations in feed rate, depth of cut, and tool geometry. Thus, if feed rate and depth of cut are increased, the cutting force will increase – resulting in a greater power requirement for metal removal. As the rake angle is increased, the cutting force decreases – resulting in a reduced power requirement for metal removal. An increase in cutting speed, although not affecting the cutting force within the range of cutting speeds normally used, would result in an increased power requirement for metal removal since power is the product of force and velocity (i.e. force × cutting speed).

It is common to relate the power required for metal cutting to the volume of metal removed in a given time, and this is referred to as power consumption. Power consumption may be given as the power in watts to remove one cubic centimetre of metal in one minute and thus has units of $W/(cm^3/min)$.

The volume of metal removed in unit time is known as the metal-removal rate and is an important aspect of machining (i.e., the greater the volume of metal removed in unit time, the more efficient is the machining process). Feed rate and depth of cut are both directly proportional to metal-removal rate; therefore, in general, it can be stated that the use of a high feed rate and a large depth of cut allows metal to be removed most efficiently in machining.

The effect of feed rate can be shown as follows. The rate at which the cutting tool will travel is the product of speed and feed; e.g. with a spindle speed of 320 rev/min and a feed of 0.15 mm/rev, the cutting tool will travel

$$320 \text{ rev/min} \times 0.15 \text{ mm/rev} = 48 \text{ mm/min}$$

If the feed rate is increased by 100% to 0.3 mm/rev and the speed is reduced by 30% to 224 rev/min (to keep tool life constant, fig. 6.8), the cutting tool will travel

244 rev/min × 0.3 mm/rev = 67.2 mm/min

This gives an increase in metal removal of

$$\frac{67.2 - 48}{48} \times 100\% = \frac{19.2}{48} \times 100\% = 40\%$$

The effect of depth of cut can be shown as follows. The volume of metal removed is the product of cutting speed, feed rate, and depth of cut; e.g. with a cutting speed of 30 m/min, a depth of cut of 2 mm, and a feed rate of 0.15 mm/rev, the volume of metal removed is

30 m/min × 2 mm × 0.15 mm/rev = 9 cm^3/min

If the depth of cut is increased by 100% to 4 mm and the cutting speed is decreased by 20% to 24 m/min (to keep tool life constant, fig. 6.9) and the feed rate is kept constant at 0.15 mm/rev, the volume of metal removed is

24 m/min × 4 mm × 0.15 mm/rev = 14.4 cm^3/min

This gives an increase in metal removal of

$$\frac{14.4 - 9}{9} \times 100\% = \frac{5.4}{9} \times 100\% = 60\%$$

However, although a large feed rate and depth of cut are beneficial for efficient metal removal, several factors limit the maximum size of cut that can be taken. These include

a) the maximum power available for cutting;
b) the maximum forces which the cutting tool can withstand;
c) the strength of the workholding system, i.e. the chuck or clamps;
d) the workpiece or cutting-tool overhang (in the case of turning and boring);
e) vibration of the machine, tool, or workpiece (resulting in 'chatter');
f) the fact that surface finish deteriorates as the size of cut is increased.

It has already been stated that an increase in cutting speed results in an increase in the power required for metal removal. However, because a greater volume of metal is removed in the same time, power consumption is found to remain constant. Thus power consumption is independent of cutting speed.

Power consumption decreases slightly with an increase in feed rate and is practically independent of depth of cut.

6.4 Determination of power consumption

Specific objective To determine experimentally the effects of variations in speed, feed, and depth of cut on power consumption.

The power supplied to a machine tool can be measured by means of a wattmeter, which is an instrument used to measure power in an electrical circuit. As previously stated, some of the total power supplied is taken up through frictional losses in the machine; the remainder is available for metal cutting. Tests of the effects of variation of cutting conditions on power consumption can best be carried out in a centre lathe equipped with a wattmeter.

To find the power required in actual cutting, it is necessary during each test to establish the power taken up in driving the machine without actually cutting, i.e. under no-load conditions. This is done by running the machine spindle and feed movement as for the required test, but without taking a cut, and recording the no-load value from the wattmeter.

This is repeated, this time taking a cut of the required depth and feed rate, and the full-load value is recorded from the wattmeter. (Multiplying factors may be necessary to convert the wattmeter reading to watts, and the manufacturer's instruction should be followed.)

Subtracting the no-load value from the full-load value gives the power required for metal cutting. This process is repeated for the number of readings required in each test.

Power consumption is the power required to remove one cubic centimetre in one minute – $W/(cm^3/min)$.

Knowing the spindle speed N rev/min and the diameter D mm of metal being machined enables the cutting speed S m/min to be calculated (from $S = \pi DN/1000$).

The volume of metal removed in cm^3/min is given by cutting speed (S m/min) × feed rate (f mm/rev) × depth of cut (d mm).

Thus power consumption $= \dfrac{\text{power required for metal cutting}}{S \times f \times d}$

The various tests can be carried out using high-speed-steel cutting tools on low-carbon steel, as follows.

Effect of variation in cutting speed

 i) Set the spindle speed to give a low cutting speed, say 20 m/min.
 ii) Set the feed rate at about 0.15 mm/rev.
 iii) With the spindle running and the feed engaged but with no cut, record the power in watts taken up by the machine transmission (the no-load power).
 iv) Set the depth of cut, say 1.5 mm, and turn a sufficient length to obtain a reading from the wattmeter (the full-load power).

Repeat this procedure for no-load and full-load conditions for about six readings, increasing the cutting speed each time.

For each cutting speed, calculate the power consumption as previously outlined.

The readings can be tabulated under the following headings:

Cutting speed (m/min) [S]	Power (W)		Power required for cutting(W) [b − a]	Vol. of metal (cm³/min) [S × f × d]	Power consumption W/(cm³/min)
	No-load [a]	Full-load [b]			

Effect of variation in feed

i) Set the spindle speed to give a cutting speed around 30 m/min.
ii) Select the feed rate at about 0.08 mm/rev.
iii) With the spindle running and the feed engaged but with no cut, record the power in watts taken up by machine transmission (the no-load power).
iv) Set the depth of cut, say 1.5 mm, and turn a sufficient length to obtain a reading from the wattmeter (the full-load power).

Repeat this procedure for no-load and full-load conditions for about six readings, increasing the feed rate each time.

For each feed rate, calculate the power consumption as previously outlined.

The readings can be tabulated under the following headings:

Feed rate (mm/rev) [f]	Power (W)		Power required for cutting(W) [b − a]	Vol. of metal (cm³/min) [S × f × d]	Power consumption W/(cm³/min)
	No-load [a]	Full-load [b]			

Effect of variation in depth of cut

i) Set the spindle speed to give a cutting speed around 30 m/min.
ii) Set the feed rate at about 0.15 mm/rev.
iii) With the spindle running and the feed engaged but with no cut, record the power in watts taken up by the machine transmission (the no-load power).
iv) Set the depth of cut to the deepest that conditions will allow and turn a sufficient length to obtain a reading from the wattmeter (the full-load power).

Repeat this procedure for full-load conditions for about six readings, decreasing the depth of cut each time.

For each depth of cut, calculate the power consumption as previously outlined.

The readings can be tabulated under the following headings:

Depth of cut (mm) [d]	Power (W)		Power required for cutting(W) [$b - a$]	Vol. of metal (cm³/min) [$S \times f \times d$]	Power consumption W(cm³/min)
	No-load [a]	Full-load [b]			

6.5 Positive and negative rake

Specific objective To explain the relative merits of positive and negative rake angles.

The rake angle of a cutting tool is the angle between the tool face (i.e. the face along which the chip slides as it is being removed from the work) and a line at right angles to the surface of the material being cut. This angle may be positive or negative as shown in fig. 6.11.

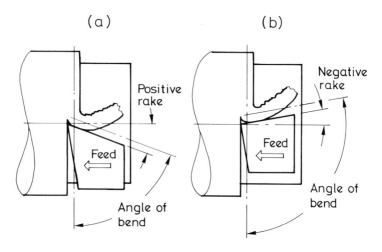

Fig. 6.11 Positive and negative rake angles

The rake angle is varied according to the strength and hardness of the material being machined, the strength of the cutting-tool material, and the amount of heat generated in cutting. Some work materials – such as plain-carbon steels and low-alloy steels – sever easily and the chip applies greatest cutting pressure at some distance from the cutting edge. In this case a large rake angle is used. Other work materials – such as cast iron and high-strength alloy steels – are more difficult to sever and the chip applies greatest pressure very close to the cutting edge. In this case a small rake angle is used.

Thus the rake angle is usually less when machining the stronger materials than when machining the weaker ones.

The stronger the cutting-tool material, the greater the rake angle can be. With large positive rake angles, the pressure of the chip on the cutting tool acts as shown in fig. 6.12(a). This can result in breakdown of the cutting edge unless the material is strong in shear. High-speed steel, for example, is strong in shear.

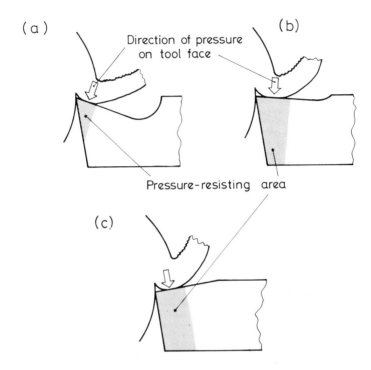

Fig. 6.12 Pressure-resisting area of different rake angles

Small positive rake angles react as shown in fig. 6.12(b).

Cemented carbides and ceramics are weak in shear but are strong in compression and, by employing a negative rake, the pressure exerted by the chip is absorbed by the tool material as shown in fig. 6.12(c).

High-strength work materials generate a great deal of heat during the cutting process and require the use of small rake angles so that greater area is available under the cutting edge to enable the heat to be dissipated.

As already stated in section 6.1, the rake angle has an effect on the cutting force and therefore on the power required in cutting. The power required in cutting increase as the rake angle is decreased, so more power is required for negative-rake tools. This is due to the way in which the chip is deformed by bending as it is removed. For example, a chip removed by a large positive rake is bent though a smaller angle than a chip removed by a negative rake, as shown in fig. 6.11. The power consumed generates heat, producing higher cutting-edge temperatures, and requires the use of cemented carbides or ceramics for negative-rake application.

Cemented carbides and ceramics are available in the form of tips (usually square or triangular) known as indexable inserts. These are held in a special holder. When one cutting edge has dulled, the insert is indexed to the next new cutting edge. This indexing is repeated until all the cutting edges have been used, and the insert is then discarded or thrown away. For this reason, they are commonly referred to as throw-away inserts.

With positive-rake inserts, the clearance angle has to be ground on the edge of the insert, fig. 6.13(a), and the cutting edges on only one face can be used.

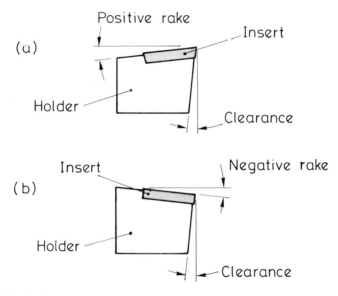

Fig. 6.13 Positive and negative inserts

With negative-rake inserts the edges are at 90° to the faces, fig. 6.13(b), and the clearance angle is obtained by the angle of seating in the tool-holder. In this case the cutting edges on both faces can be used, i.e. there are twice as many as with a positive-rake insert.

In many instances the use of a negative rake is purely for the economic advantage of having twice the number of cutting edges than are available on a positive-rake insert.

In general, negative-rake tools are used

a) for general-purpose machining of most materials, especially rough or interrupted cuts;

b) for tough materials on rigid set-ups;
c) for greatest economy, since inserts can be turned over.

In general, positive-rake tools are used

a) for machining softer steels and non-ferrous metals;
b) for slender parts which will not withstand high cutting forces;
c) on low-powered machines or set-ups which lack rigidity.

6.6 Cutting-tool materials

Specific objective   To outline the advantages and limitations of cutting-tool materials.

To be effective, the material from which a cutting tool is made must possess certain properties, the most important of which are red hardness, abrasion resistance, and toughness.

Red hardness   It is obvious that a cutting tool must be harder than the material being cut, otherwise it will not cut. It is equally important that the cutting tool remains hard even when cutting at high temperatures. The ability of a cutting tool to retain its hardness at high cutting temperatures is known as red hardness.

Abrasion resistance   When cutting, the edge of a cutting tool operates under intense pressure and will wear due to abrasion by the material being cut. Basically, the harder the material the better its resistance to abrasion.

Toughness   A cutting-tool material which is extremely hard is unfortunately also brittle. This means that a cutting edge will chip on impact if, for example, the component being machined has a series of slots and the cut is therefore intermittent. To prevent the cutting edge from chipping under such conditions, it is necessary that the material has a certain amount of toughness. This can be achieved only at the expense of hardness; that is, as the toughness is increased so the hardness decreases.

It can be readily seen that no one cutting-tool material will satisfy all conditions at one time. A cutting tool required to be tough due to cutting conditions will not be at its maximum hardness and therefore not be capable of fully resisting abrasion. Alternatively, a cutting tool requiring maximum hardness will have maximum abrasion resistance but will not be tough enough to resist impact loads. The choice of cutting-tool material is governed by the type of material to be cut and the conditions under which cutting is to take place, as well as the cost of the tool itself. Remember that cutting tools are expensive, and great care should be taken to avoid damage and consequent wastage both in use and during resharpening.

High-carbon steel

High-carbon steel is an alloy of iron and carbon (between 1% and 1.4%). In a hardened and tempered condition it can be used as a cutting-tool material. However, it begins to lose its hardness at low temperatures (around 200°C) and so is limited to use at low cutting speeds, e.g. for taps, dies, and cold chisels, and on soft non-ferrous metals which give low cutting temperatures.

High-speed steels (HSS)

High-speed tool steel consist of iron and carbon with differing amounts of alloying elements such as tungsten, chromium, vanadium, and cobalt. When hardened, these steels are brittle and the cutting edge will chip on impact or with rough handling. They have a high resistance to abrasion but are not tough enough to withstand high shock loads. These steels will cut at high speeds and will retain their hardness even when the cutting edge is operating at temperatures around 600°C.

A general-purpose high-speed tool steel used to manufacture drills, reamers, taps, milling cutters, and similar cutting tools contains 18% tungsten, 4% chromium, and 1% vanadium and is referred to as an 18–4–1 tool steel. The addition of 8% cobalt to the above high-speed steel increases the hardness and red hardness and produces a steel which can be used at higher speeds. These tool steels are often referred to as super-high-speed steels.

Apart from being used to manufacture the cutting tools already mentioned, high-speed steel is available as 'tool bits' in round, square, or rectangular section already hardened and tempered. The operator has only to grind the required shape on the end before using.

To save on cost, cutting tools such as lathe tools are made in two parts, instead of from a solid piece of expensive high-speed steel. The cutting edge at the front is high-speed steel and this is butt-welded to a tough steel shank. These tools are known as butt-welded tools, fig. 6.14. The cutting edge can be reground until the high-speed steel is completely used and the toughened shank is reached. At this stage the tool is thrown away.

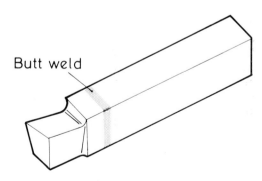

Fig. 6.14 Butt-welded lathe tool

Stellite

Stellite is a cobalt–chromium–tungsten alloy containing no iron. It cannot be rolled or forged and is shaped by casting, from which it derives its cutting properties and hardness. No other form of heat treatment is required. Stellite is as hard as high-speed steel and has a higher red hardness, retaining its hardness at temperatures of 700°C. Being cast, it is also brittle and care must be taken to avoid chipping the cutting edge. It is more expensive than high-speed steel and is supplied as solid cast tool bits of round, square, and rectangular section, as tips, and as tipped tools where the tip is brazed to a toughened steel shank.

Stellite can be reground using standard grinding wheels, but care must be taken to avoid overheating, which leads to surface cracking and subsequent breakdown of the cutting edge.

Cemented carbides

Cemented carbides are produced by a powder-metallurgy technique, i.e. by using metals in their powder form. The final mixture of powders consists of various amounts of hard particles and a binding metal. The hard particles give the material its hardness and abrasion resistance, while the binding metal provides the toughness.

The most common hard particle used is tungsten carbide, but titanium carbide and tantalum carbide are often added in varying amounts. The binding metal used is cobalt, and various grades of cemented carbide are obtained for cutting different groups of materials by mixing in different proportions.

Cemented carbides normally contain 70% to 90% of hard particles together with 10% to 30% cobalt binding metal. In general, the more cobalt that is present, the tougher the cemented carbide. Unfortunately, however, this increase in toughness obtained by increasing the cobalt content results in decreased hardness and abrasion resistance.

Cemented carbides are used in cutting tools for turning, milling, drilling, boring, etc. in the form of tips or inserts which are brazed or clamped to a suitable tool shank, fig. 6.15. The blanks are produced by mixing the metal powders in the correct proportions, pressing them into the required shape, and finally heating at temperatures as high as 1600°C, a process known as sintering. This sintering stage results in the cobalt binding metal melting and fusing with the hard particles, or cementing, to form a solid mass – hence the name cemented carbides.

Cemented carbides are classified into three main groups: those used for machining steel, designated by the letter P and coloured blue; those used for machining cast iron and non-ferrous metals, designated K and coloured red; and finally multi-purpose grades, designated M and coloured yellow. These letters are followed by a number which, as it increases, denotes increasing toughness with a resultant decrease in hardness.

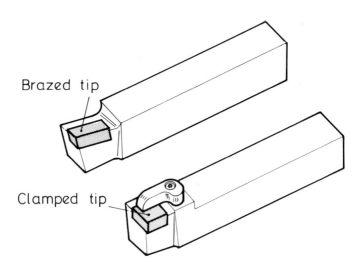

Fig. 6.15 Tipped lathe tools

Cemented carbides have a red hardness higher than both high-speed steel and stellite and will retain their hardness at temperatures well in excess of 700°C.

Cemented-carbide tips are available in which a thin layer of titanium nitride is bonded over all the surfaces. This coating is extremely hard and has a low coefficient of friction, leading to an increase in abrasion resistance and a longer-lasting cutting edge. It is claimed that speeds can be increased by 50% above those used with conventional cemented carbides with the cutting edge lasting the same time, or, alternatively, the cutting edge will last twice as long as that of the conventional tip if run at the same speed.

Owing to their extreme hardness, cemented carbides cannot be reground using the same wheels used to regrind high-speed steel and stellite. Silicon-carbide wheels (usually green in colour) must be used – these maintain a keen cutting action. To finish the cutting edge it is necessary to use a diamond wheel which laps the surfaces to produce a keen edge. Great care must be taken to avoid overheating, which leads to surface cracking of the tip and subsequent breakdown of the cutting edge.

Brazed-tip tools are expensive and have the disadvantages that they have to be removed from the machine to be reground and must then be reset in the machine, and after each regrind the tip becomes smaller and smaller. Tips which are clamped in a suitable holder – known as throw-away tips or inserts – do not suffer from these disadvantages. When one cutting edge is worn, the insert is merely unclamped, turned to the next keen cutting edge, and reclamped. This is repeated until all cutting edges are used (there may be as many as eight) and the insert is then thrown away. The holder is not removed from the machine, and no resetting is necessary. Both brazed-tip and insert types do, however, have their applications in industry.

Ceramic cutting materials
Two types of ceramic cutting materials are available: a material made from pure aluminium oxide and a mixed ceramic of titanium carbide and aluminium oxide. The latter is used to cut the higher-strength steels and chilled cast iron. The ceramic powders are mixed, pressed into the required shape, and finally sintered, resulting in a solid dense blank which is subsequently ground to the correct size.

The tip blanks are used by clamping to a suitable tool holder.

Ceramic cutting materials have a high abrasion resistance and high red hardness. They show no deformation even at temperatures up to 1000°C, remaining hard at temperatures which would affect cemented carbides. They can be used to cut cast iron, spheroidal-graphite cast iron, malleable iron, and alloy steels at cutting speeds from 100 to 600 m/min at cutting depths of up to 6 mm in cast iron.

Use of a cutting fluid is not recommended because of the danger of thermal shock – pure aluminium oxide will be destroyed by a sudden temperature change of more than 200°C.

Cubic boron nitride
Next to diamond, cubic boron nitride is the hardest known material, with exceptionally high abrasion resistance and cutting-edge life in severe cutting conditions.

A layer of cubic boron nitride approximately 0.5 mm thick is bonded to a cemented-carbide tip approximately 5 mm thick. The cemented carbide provides a shock-resisting base. This material will machine chilled cast iron and fully hardened steel and still maintain a cutting edge. It is designed to perform most effectively on materials difficult to cut, and in some cases can replace a grinding operation.

It does not react with other metals or oxidise at temperatures below 1000°C and it is therefore virtually unaffected by heat generated in the high-speed cutting of difficult-to-machine materials.

Diamond
Diamond is the hardest known material. For this reason, single-crystal natural-diamond tools have been used in industry for a great number of years, to dress grinding wheels and as cutting tools to finish-machine non-ferrous and non-metallic materials.

A synthetic or man-made diamond material is now available which is extremely tough, with a hardness approaching that of natural diamond. A layer of this synthetic diamond material is bonded to a tough shock-resisting base of cemented carbide for use in the form of tips. The range of application is on non-ferrous metals such as aluminium alloys, magnesium alloys, copper, brass, bronze, and zinc alloys and non-metallic materials such as ceramics, porcelain, and plastics. This material will also machine fully sintered tungsten carbide.

High-cutting speeds are employed – up to 1200 m/min on non-ferrous metals, while sintered tungsten carbide can be machined at between 150 and 500 m/min depending on the conditions. Under the same cutting conditions on the materials listed above, this synthetic diamond material will outlast all other cutting-tool materials, including single-crystal natural diamond, which is prone to chipping at the cutting edge.

6.7 Cutting fluids

Specific objective   To understand the function of cutting fluids.

During metal cutting, the metal immediately ahead of the cutting tool is severely compressed, which results in heat being generated. The metal then slides along the tool face, friction between the two surfaces generating additional heat. Any rubbing between the tool and the cut surface, which would occur with tool wear when the clearance angle is reduced, also produces heat. This heat is usually detrimental, especially to high-speed-steel cutting tools. Some metals, as they are cut, have a tendency to produce a chip which sticks or welds to the tool face, due chiefly to the high pressure between the metal and the tool. This has the effect of increasing the power required for cutting, increasing friction and therefore heat and finally, as the chip breaks away from the tool face and reforms, it creates wear on the tool face and a bad surface finish on the work. Excessive heat generated during cutting may be sufficient to cause the work to expand. Work measured under these conditions may be undersize when it cools.

The basic role of a cutting fluid is to control heat, and it may do this by direct cooling of the work, chip, and tool or by reducing friction by lubricating between the work, chip, and tool. To cool effectively, a cutting fluid should have a high specific heat capacity and a high thermal conductivity.

The fluids most readily associated with cooling and lubricating are water and oil. Water has a higher specific heat capacity and thermal conductivity than oil, but unfortunately will promote rust and has no lubricating properties. Oil does not promote rust, has good lubricating properties, but does not cool as well as water. To benefit from the advantages of each, they can be mixed together with various additives to give a required measure of cooling and lubrication. With the high cost of oil, the cost savings of water-based fluids are so great that a great deal of development is being carried out to provide fluids which have good lubricating properties when mixed with water.

In general, the use of cutting fluids can result in

a) less wear on cutting tools,
b) the use of higher cutting speeds and feeds,
c) improved surface finish,

d) reduced power consumption,
e) improved control of dimensional accuracy.

The ideal cutting fluid, in achieving the above, should

i) not corrode the work or machine,
ii) have a low evaporation rate,
iii) be stable and not foam or fume,
iv) not injure or irritate the operator.

6.8 Types of cutting fluid

Specific objective To select and give reasons for the use of a suitable cutting fluid for a particular machining operation.

Neat cutting oils

These oils are neat in so much as they are not mixed with water for the cutting operation. They are usually a blend of a number of different types of mineral oil, together with additives for extreme-pressure applications. Neat cutting oils are used where severe cutting conditions exist, usually when slow speeds and feeds are used or with extremely tough and difficult-to-machine steels. These conditions require lubrication beyond that which can be achieved with soluble oils.

In some cases soluble oil cannot be used, due to the risk of water mixing with the hydraulic fluid or the lubricating oil of the machine. A neat oil compatible with those of the machine hydraulic or lubricating systems can be used without risk of contamination. Neat cutting oils do not have good cooling properties and it is therefore more difficult to maintain good dimensional accuracy. They are also responsible for dirty and hazardous work areas by seeping from the machine and dripping from workpieces and absorbing dust and grit from the atmosphere.

Low-viscosity or thin oils tend to smoke or fume during the cutting operation, and under some conditions are a fire risk.

The main advantages of neat cutting oils are their excellent lubricating property and good rust control. Some types do, however, stain non-ferrous metals.

Soluble oils

Water is the cheapest cooling medium, but it is unsuitable by itself, mainly because it rusts ferrous metals. In soluble oils, or more correctly emulsifiable oils, the excellent cooling property of water is combined with the lubricating and protective qualities of mineral oil. Oil is, of course, not soluble in water, but with the aid of an agent known as an emulsifier it can be broken down and dispersed as fine particles throughout the water to form an emulsion.

Other ingredients are mixed with the oil to give better protection against corrosion, resistance to foaming and attack by bacteria, and prevention of skin irritations. Under severe cutting conditions where cutting forces are high, extreme-pressure (E.P.) additives are incorporated which do not break down under these extreme conditions but prevent the chip welding to the tool face.

Emulsions must be correctly mixed, otherwise the result is a slimy mess. Having selected the correct ratio of oil to water, the required volume of water is measured into a clean tank or bucket and the appropriate measured volume of soluble oil is added gradually at the same time as the water is slowly agitated. This will result in a stable oil/water emulsion ready for immediate use.

At dilutions between 1 in 20 and 1 in 25 (i.e. 1 part oil in 20 parts water) the emulsion is milky white and is used as a general-purpose cutting fluid for capstan and centre lathes, drilling, milling, and sawing.

At dilutions from 1 in 60 to 1 in 80 the emulsion has a translucent appearance, rather than an opaque milky look, and is used for grinding operations.

For severe cutting operations, such as gear cutting or broaching and machining tough steels, fluids with E.P. additives are used at dilutions from 1 in 5 to 1 in 15.

As can readily be seen from the above, when the main requirement is direct cooling, as in the case of grinding, the dilution is greater, i.e. 1 in 80. When lubrication is the main requirement, as with gear cutting, the dilution is less, i.e. 1 in 5.

The advantages of soluble oils over neat cutting oils are their greater cooling capacity, lower cost, reduced smoke, and elimination of fire hazard. Disadvantages of soluble oils compared with neat cutting oils are their poorer rust control and that the emulsion can separate, be affected by bacteria, and become rancid.

Synthetic fluids
Sometimes called chemical solutions, these fluids contain no oil but are a mixture of chemicals dissolved in water to give lubricating and anti-corrosion properties. They form a clear transparent solution with water, and are sometimes artificially coloured. They are very useful in grinding operations, where, being non-oily, they minimise clogging of the grinding wheel and are used at dilutions up to 1 in 80. As they are transparent, the operator can see the work, which is also important during grinding operations.

They are easily mixed with water and do not smoke during cutting. No slippery film is left on the work, machine, or floor. They give excellent rust control and do not go rancid. At dilutions of between 1 in 20 and 1 in 30 they can be used for general machining.

Semi-synthetic fluids
These are recently developed cutting fluids, sometimes referred to as chemical emulsions. Unlike synthetic fluids, these fluids do have a small amount of oil emulsified in water, as well as dissolved chemicals, but they

are not true emulsions. When mixed with water they form extremely stable transparent fluids, with the oil in very small droplets. Like the synthetic types, they are often artificially coloured for easy recognition.

They have the advantage over soluble oil of increased control of rust and rancidity and a greater application range. They are safer to use, will not smoke, and leave no slippery film on work, machine, or floor. Depending on the application, the dilution varies between 1 in 20 and 1 in 100.

6.9 Application of cutting fluids

Specific objective To recognise the need for correct application of cutting fluids.

Having selected the correct type of cutting fluid, it is equally important to apply it correctly. This is best done by providing a generous flow at low pressure to flood the work area. Flooding has the added advantage of washing away the chips produced. Fluid fed at high pressure is not recommended, since it breaks into a fine spray or mist and fails to cool or lubricate the cutting zone. To cope with the large flow of fluid, the machines must have adequate splash guards, otherwise the operator tends to reduce the flow and the resulting dribble does little to improve cutting.

Many methods have been used to direct the fluid into the cutting zone and from every possible direction. The shape of the nozzle is important but depends largely on the operation being carried out and on the shape of the workpiece. The nozzle may be a simple large-bore pipe or be flattened as a fan shape to provide a longer stream. The main flow may be split into a number of streams directed in different directions – up, down, or from the sides – or, by means of holes drilled in a length of pipe, create a cascade effect. In some cases – especially with grinding, where the wheel speed creates air currents which deflect the cutting fluid – deflector plates are fitted to the pipe outlet. Where the cutting tool is vertical, it can be surrounded by a pipe having a series of holes drilled into the bore and directed towards the cutting tool. Whatever the method used, the fundamental need is to deliver continuously an adequate amount of cutting fluid where it is required.

There are machining operations where difficulties do arise. One such case is drilling. It is extremely difficult, if not impossible, to get fluid down the flutes of a drill which are full of metal chips coming up. It is necessary to withdraw the drill frequently to allow fluid to reach the cutting edge. This method does not ensure a continuous supply of fluid, and the frequent withdrawal of the drill takes time and adds to the cost of the operation. This is especially true when drilling deep holes. To overcome this problem, drills are available with two internal oil holes which follow the spiral of the drill and emerge on the lips of the drill, fig. 6.16. The cutting fluid is fed by pipe through a collar which is held stationary by a stop on the machine. The fluid then flows through each hole to the cutting edges

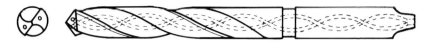

Fig. 6.16 Oil-feed drill

and back along the flutes. Metal chips are assisted along the flutes by the fluid, eliminating the need for frequent withdrawal.

6.10 Safety in the use of cutting fluids

Specific objective   To appreciate precautions to be observed when using cutting fluids.

Cutting fluids can affect the health of those exposed to them in various ways: by contact with the skin, by contact with the eyes, if they are breathed in with the air as small droplets or vapour, or if they are swallowed. The possible effects of these can normally be avoided by good housekeeping and a high standard of personal hygiene.

The following precautions, if observed, will reduce or eliminate the likely hazards.

i) Working methods should be employed that avoid direct skin contact with oils, e.g. machine splash guards and correct handling procedures should be used.
ii) Adequate local exhaust ventilation should be provided for areas where vapour and mists are generated.
iii) Adequate protective clothing should be worn.
iv) Only disposable 'wipes' or clean rags should be used. Contaminated rags and tools should never be put into overall pockets.
v) A barrier cream should be applied to the hands and exposed areas of the arms before starting work and on resuming work after a break.
vi) Hands should be thoroughly washed using suitable hand cleaners and warm water and dried using a clean towel before, as well as after, going to the toilet, before eating, and at the end of each shift.
vii) Conditioning cream, applied after washing, replaces fatty matter in the skin and helps prevent dryness.
viii) Contaminated clothing, especially undergarments, should be changed regularly and be thoroughly cleaned before reuse.
ix) Overalls should be cleaned frequently.
x) Paraffin, petrol, and similar solvents should not be used for skin-cleansing purposes.
xi) All cuts and scratches must receive prompt medical attention.
xii) You should seek prompt medical advice if you notice any skin abnormality.

Exercises on chapter 6

1 Why is the use of negative-rake inserts more economical than the use of positive-rake inserts?
2 List five precautions which should be observed when using cutting fluids.
3 Why is tool life an important factor in metal cutting?
4 State the three basic types of chip formation and state the conditions under which each may be produced.
5 State two main advantages of using a neat cutting oil.
6 Describe and name two ways in which the major cutting edge of a single-point cutting tool can be presented to the workpiece.
7 What are the conditions to be adopted in order to give the highest volume of metal removal during a given tool life?
8 State four factors which limit the maximum size of cut which can be taken in a metal-cutting operation.
9 What factor has the biggest effect on tool life, and why?
10 State a cutting application for which cubic boron nitride is best suited.

7 Machine tools

General objective The student explains the basic principles of operation and the uses of machine tools.

A machine tool may be described as a power-driven device capable of supporting and securing a workpiece and cutting tool and of providing the necessary motions for metal removal in order to give the required shape to the workpiece surface.

In order to obtain a finished workpiece of the correct shape and size, surplus material must be removed in the form of chips as it is machined in the machine tool.

The shape of the machined surface that is produced depends on the motions imparted by the machine tool to the workpiece and the cutting tool, the relationship between these motions, and the shape of the cutting tool. Thus, by varying the relative motions and by changing the shape of the cutting tool, it is possible to obtain a variety of shapes using the same machine tool.

For metal removal to take place, there must be a relative movement between the cutting tool and the workpiece so that any material coming in the path of the cutting tool is removed in the form of a chip. This movement is known as the primary motion and may, depending on the machine tool, be given to the cutting tool or to the workpiece. This is shown as P in fig. 7.1.

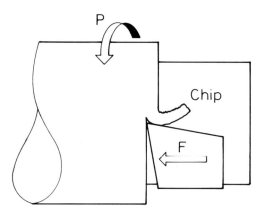

Fig. 7.1 Relative movements for metal cutting

Once the material in the path of the cutting tool has been removed, no further material can be removed until either the cutting tool or the workpiece is moved to extend the cutting process to the complete surface to be machined. This second motion – known as the feed motion – may be continuous as in turning operations or intermittent as in shaping machines. This is shown as F in fig. 7.1.

In order to advance the cutting tool to the surface of the work and subsequently to withdraw it, a third motion is required. This motion determines the depth of cut and controls the size of the workpiece, e.g. the diameter in a turning operation.

The most common types of primary motion are rotation and straight-line reciprocating motion, i.e. movement backwards and forwards. Rotary motion may be transmitted either to the workpiece, as in turning, or to the cutting tool as in milling, drilling, and grinding.

Straight-line reciprocating motion is employed in surface grinders, shaping machines, and other machines such as planing, slotting, and broaching machines and power hacksaws. In the case of shaping machines, this motion is transmitted to the cutting tool.

The standard machine tools used in industry are designed such that their motions or combination of motions will produce the desired workpiece shape. Thus it can be seen that the range of standard machine tools can be used to produce a variety of shapes or combination of shapes, such as cylinders – both external (shafts) or internal (holes) – cones (tapers), plane surfaces, and spheres or parts of them.

7.1 Generating, forming, and copying

Specific objective   To explain how shapes are produced by generating, forming, and copying, or by a combination of these.

As previously stated, the shape of a machined surface depends on the motions imparted by the machine tool to the workpiece and the cutting tool, the relationship between these motions, and the shape of the cutting tool. Depending on these factors, the machined surface is produced by generating, forming, or copying, or by a combination of these.

In generating, the surface produced is the result of a large number of successive relative motions between the workpiece and the cutting tool, the accuracy being dependent on the accuracy of the machine-tool motions and not on the shape of the cutting tool. For example, in turning a parallel diameter of a certain length, the workpiece is rotated about its axis as the cutting tool is moved in a straight line parallel to the axis. The accuracy of the resulting cylinder is dependent on that of the primary motion, i.e. the spindle rotation, for its roundness and on that of the feed motion for parallelism. The accuracy of the cylindrical shape is independent of the shape of the cutting tool. If the feed motion is in a straight line at an angle to the axis of rotation, a taper cylinder or cone will be generated.

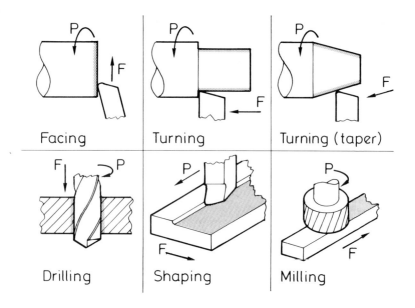

Fig. 7.2 Examples of generated surfaces

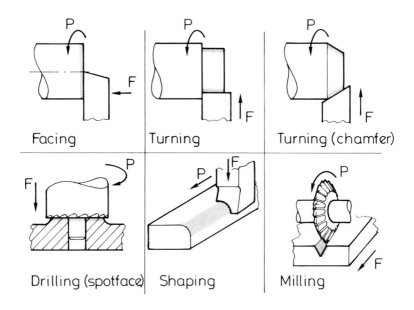

Fig. 7.3 Examples of formed surfaces

A flat surface can be generated in turning by feeding the tool in a straight line at right angles to the axis of rotation, e.g. when facing the end of a workpiece.

Some examples of surfaces generated on various machine tools are shown in fig. 7.2.

In forming, the surface produced is a mirror image of the shape of the cutting tool, the accuracy being a function of the accuracy of form of the cutting tool. Some examples of surfaces formed on various machine tools are shown in fig. 7.3.

In some cases the surface is produced by a combination of both generating and forming. For example, in milling a flat surface using a cylindrical cutter, the shape across the surface is dependent on the cutter shape while the accuracy along its length is dependent on the feed motion, fig. 7.4(a). Similarly in screw cutting: the accuracy of the thread profile is dependent on the form tool while the accuracy of pitch is dependent on the co-ordinated movement of the primary and feed motions, fig. 7.4(b).

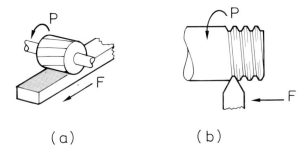

Fig. 7.4 Combination of generating and forming

In copying, the cutting tool is fed along a path controlled by the profile of a template or model so that the required shape is reproduced on the workpiece. In this case, two co-ordinated feed motions are applied simultaneously, as shown on the copy-turned profile in fig. 7.5.

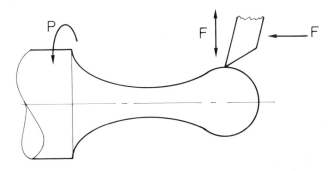

Fig. 7.5 Surface produced by copying

7.2 Power transmission of machine tools

Specific objective To draw simple block diagrams to describe the basic elements of power transmission from the input to the work and the cutting tool.

The power input to a machine tool is obtained from an electric motor. The power from the motor may be used to drive a pump to supply fluid to cylinders through a series of valves in the case of hydraulically operated machines, or it may be transmitted by belt drive to a gearbox and through a clutch or gear-change system for mechanical operation, or it may simply be transmitted by belt drive alone. Separate power-transmission systems may be arranged for primary motion and for feed motion or they may be interconnected.

Drilling machines

The spindle of a small sensitive drilling machine is driven from an electric motor by a belt, through stepped pulleys to give a range of speeds. Speed change is effected by shifting the belt to the appropriate diameter of pulley. In addition, some machines have a small two-speed gearbox, thus doubling the number of available speeds. These small machines do not have power feed to the quill, which is hand-operated to give a greater degree of sensitivity.

Heavy-duty pillar drilling machines have a gearbox driven from the electric motor, spindle speeds being selected by means of a lever connected to change gears. A drive from the gearbox to a feed gearbox provides a limited range of feeds to the quill – this is necessary when drilling larger-diameter holes. A lever is provided to engage and disengage this power feed. A block diagram of this arrangement is shown in fig. 7.6.

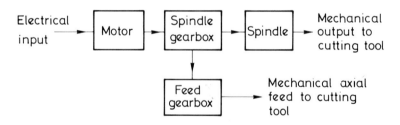

Fig. 7.6 Power transmission of a drilling machine

Shaping machines

In shaping machines, power is transmitted from an electric motor to a gearbox by belt drive. The bull gear, which is connected to the rocking bracket to provide the reciprocating motion to the ram, is driven from the

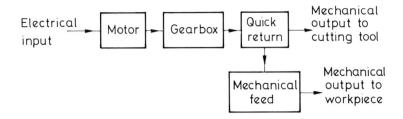

Fig. 7.7 Power transmission of a shaping machine

gearbox via a clutch controlling the machine stop and start. Power feed of the cross slide is by means of a mechanical link through a gear driven by another gear at the end of the bull-gear spindle. A block diagram of this arrangement is shown in fig. 7.7.

Milling machines
Power transmission in milling machines is again from an electric motor by belt drive to a gearbox and through a series of gears to the machine spindle. A lever is used to shift gears to give a range of spindle speeds. The gearbox drives a feed gearbox to provide power feed to the knee, saddle, and table motions. Levers are used to select the rate of power feed for cutting or for rapid traverse motion when setting up. The drive from the main gearbox to the feed gearbox, contained in the knee, is by a telescopic shaft universally jointed. On large machines, a separate electric motor mounted in the knee provides power to the feed gearbox. Adjustable trip dogs are provided to disengage the feed movements at any point within the specified traverse range. A block diagram of a typical arrangement is shown in fig. 7.8.

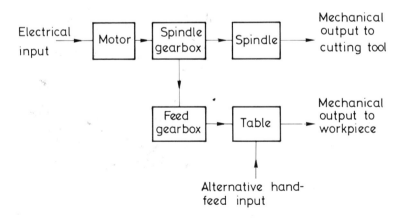

Fig. 7.8 Power transmission of a milling machine

Surface-grinding machines

In surface-grinding machines, the spindle is driven from an electric motor through a belt. Although hand operation is provided, the reciprocating table and cross-slide movements are hydraulically operated. Power is from an electric motor to a pump. Hydraulic fluid is delivered to a cylinder for each movement, through a series of valves to control rate of traverse and direction. A block diagram of a typical arrangement is shown in fig. 7.9.

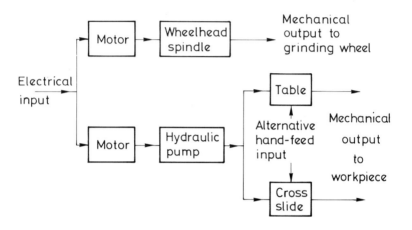

Fig. 7.9 Power transmission of a surface-grinding machine

Capstan and turret lathes

These machine tools are used for quantity-production work where the number of parts to be produced is too great to be done economically on a centre lathe and insufficient to be done economically on an automatic lathe.

The major difference between these machines and a centre lathe is the addition of a six-sided turret capable of holding a number of cutting tools which can be rapidly indexed to bring the required tool into the desired position for cutting. Turret indexing is done automatically at the end of each withdrawal stroke.

The difference between a capstan and a turret lathe is in the design of the turret slide. In a turret lathe, fig. 7.10, the hexagonal tool head is mounted directly on the machine bed and is capable of moving along the entire length of the bed, making it more suitable for larger work. The hexagonal tool head of the capstan lathe, fig. 7.11, has a short slide mounted on a base which can be clamped to the machine bed at any desired position along its length. This is positioned during setting of the machine, and the restricted length of slide makes the capstan lathe more suitable for short-length work. The turret slide on both capstan and turret lathes is provided

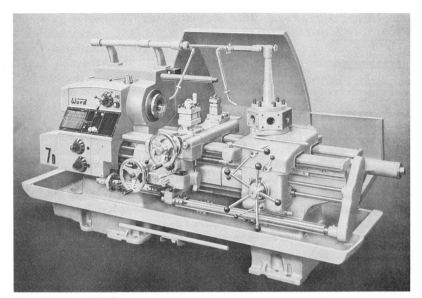

Fig. 7.10 Turret lathe

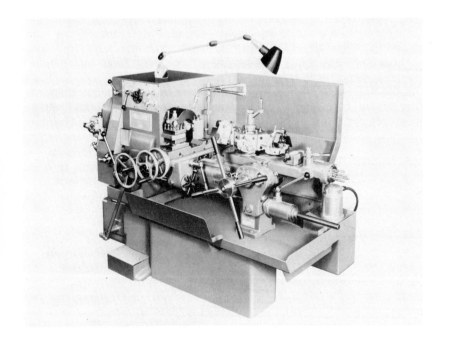

Fig. 7.11 Capstan lathe

with a limited range of feeds which are automatically tripped by a stop. A separate stop is provided for each turret face.

The headstock of capstan and turret lathes differs from that of a centre lathe in that spindle speeds can be rapidly changed without the need to stop the spindle. In some cases, the next required speed can be preselected while machining is being carried out and, when required, is brought into operation merely by the pushing of a button. The drive to the gearbox contained in the headstock is by belt drive from an electric motor. The gears are in constant mesh, and the speed changes are effected through a series of clutches and are carried out while the spindle is rotating. A separate drive is taken to a feed gearbox which supplies power for a limited number of feed rates to the turret and in some cases to the saddle movement.

In most cases, the saddle is moved by hand to the required position against a stop and is locked in position. The cross slide is then operated in or out.

The cross slide carries an indexable four-way toolpost at the front in which cutting tools are mounted for the production of undercuts, chamfers, etc. A rear toolpost is also provided and usually carries the parting-off tool for use with bar work. The cross slide is also equipped with stops for each direction.

The smaller capstan lathes are usually confined to work produced from bar held in a collet chuck. The collet can be opened, the bar be automatically fed out the required distance against a stop mounted in the turret, and the collet be closed while the spindle is running. Collets are available to hold a variety of shapes – round, square, rectangular, and hexagonal. The collet chuck may be hand- or power-operated. Other methods of workholding such as jaw chucks, both hand- and power-operated, as well as fixtures are available.

A block diagram of the power transmission of a capstan/turret lathe is shown in fig. 7.12. Apart from the simple operations carried out from the cross slide as previously outlined, all operations are carried out from the

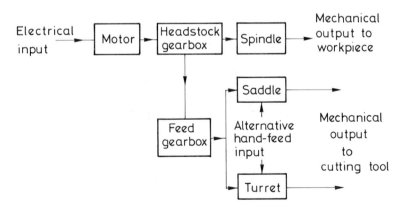

Fig. 7.12 Power transmission of a capstan/turret lathe

turret, taking advantage of the number of tool stations and of the use of combined machining operations. A wide range of tooling is available for use from the turret, some examples of which are described here.

Adjustable stop   An adjustable stop is used when producing workpieces from bar. The stop is either fitted directly in the bore of the turret and held by means of a clamping bolt or is fitted in a holder fitted to the front of the turret face.

At the start of the machining operation, the bar is fed out against the stop, which is adjusted to ensure that the correct length of bar protrudes. This is shown in fig. 7.13.

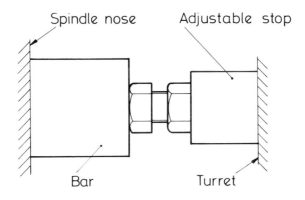

Fig. 7.13 Adjustable stop

Boring-bar holder   This is fitted to the turret face to carry boring-bars and other shank-mounted tools, such as an adjustable stop, fig. 7.14. Split adapter bushes to suit different sizes of shank are available and fit in the bore of the holder.

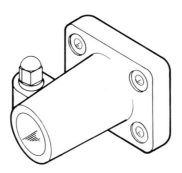

Fig. 7.14 Boring-bar holder

Flat centring tool   This tool is used to prepare a true start for a drill when a drilling operation has to be performed. It consists of a rigid flat blade, usually having an included angle of 120°, located in a bush, fig. 7.15, which is held in the bore of a boring-bar holder.

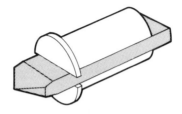

Fig. 7.15 Flat centring tool

Centring-and-facing tool   This tool is also used to prepare a true start for a drill and includes flat blade cutters which may be arranged for plain facing, counterboring, or chamfering, fig. 7.16. Thus the end of the workpiece is machined simultaneously with the centring operation.

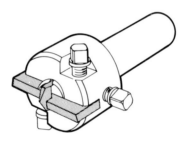

Fig. 7.16 Centring-and-facing tool

Knee turning toolholder   This toolholder is used for turning operations and mounts directly on the turret face. One or more cutting tools may be accommodated in a holder and can be adjusted to cut the correct diameter. A drill or boring-bar can be fitted in the central bore, thus enabling turning and drilling or boring to be carried out simultaneously, fig. 7.17. A supporting bush is fitted which locates with a support bar fitted to the headstock of the capstan lathe to ensure complete rigidity when taking roughing cuts.

Roller-steady box tool   This type of tool is used to turn plain diameters. Those capable of turning up to 25 mm diameter fit in the bore of the turret, and those of larger capacity fit directly on the turret face.

Fig. 7.17 Knee turning toolholder

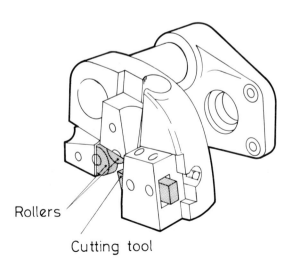

Fig. 7.18 Roller-steady box tool

The box tool shown in fig. 7.18 carries a cutting tool and a pair of roller steadies which enable a highly finished diameter to be produced in one cut at high speed and with a fast feed. A highly burnished finish is produced by the rollers, due to the metal being pushed against the rollers by the force of the cutting action. In view of this, it is usually the case that a deeper cut produces a better finish.

The rollers are mounted on slides which are adjusted by screws to suit the required diameter of the workpiece, and the cutting tool is adjustable in its holder.

The box tool, of rigid box-type design, is capable of taking heavy cuts, using carbide or high-speed-steel cutting tools.

Self-opening diehead This type of diehead is employed to produce external threads on a workpiece. Solid dies require spindle reversal to remove them on completion of the operation. The self-opening types release the dies outwards, clear of the thread, when the operation is complete and can be quickly retracted without the need to reverse the spindle. The dies can then be quickly closed in readiness for the next threading operation.

Numerous types of self-opening diehead are available, those with radial dies being common and generally referred to as Coventry dieheads, fig. 7.19(a). Four dies made from high-speed steel, shown in fig. 7.19(b), are located in slots in the inner body and can be made to move radially in

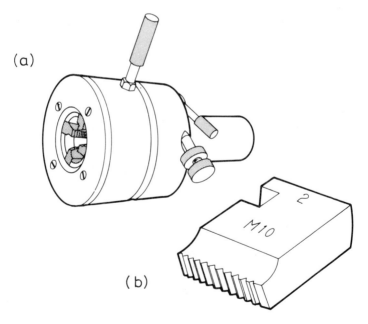

Fig. 7.19 Self-opening diehead and die

and out by means of a cam on the inner surface of the partially rotatable outer body.

In operation, the turret moves forward until it reaches its stop. The dies and outer body keep moving forward under the action of the screw thread being produced. The outer body is then pulled out of engagement with an indent pin and partially rotates, pulling the dies outwards into an open position clear of the thread. Die closing is effected by partially rotating the outer body in the opposite direction, by hand, until the indent pin re-engages, the inner cam pushing the dies to their inward position.

Dieheads are adjustable to maintain correct-size threads and are provided with an arrangement which allows a roughing cut and a finishing cut to be taken successively by the same diehead, a feature which is especially required with coarse-pitch threads.

Speeds for threading with dieheads are generally one fifth to one sixth of the speed required for turning operations using high-speed-steel cutting tools.

7.3 Control of dimension and form

Specific objective To describe slideway systems, screw and nut movements, and calibration dials for control of dimension and form.

As already stated, the shape of a machined surface depends on the motions imparted by the machine tool to the workpiece and the cutting tool and on the relationship between these motions.

Any body in space possesses six degrees of freedom, as shown in fig. 7.20. This body may have linear motion along any one of the three

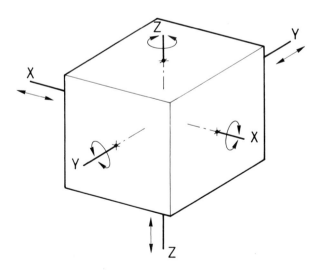

Fig. 7.20 Six degrees of freedom

mutually perpendicular axes X, Y, or Z or it may have rotational motion about any one of these three axes. The branch of science relating to problems of motion is known as kinematics. The aim of kinematic design is to prevent movement in certain degrees of freedom by applying constraints.

The design of a machine tool is such that each element is constrained from moving in any direction other than the one required; for example, a machine spindle is constrained so that it will rotate about only one axis and have no linear motion. Thus a machine tool must provide for linear motion by means of slides and for rotational motion by means of spindles all accurately aligned relative to each other, to give control of form together with accurate means of control of dimension.

To produce parts of accurate form, slide systems of machine tools must move in a straight line and be aligned to give parallelism and squareness with other motions. All unnecessary degrees of freedom must be constrained, and the slide must move easily in the direction required, free from unnecessary restraint. Provision must be made for adjustment to compensate for wear, for adequate lubrication, and for prevention of swarf accumulation and for swarf removal.

Common basic elements of slide design are cylindrical, flat, vee, and dovetail slides, which may be used separately or combined in various ways.

Cylindrical slides are incorporated in simple forms of pillar drilling machines and radial-arm drilling machines, the design providing two degrees of freedom: a linear motion along the axis and a rotary motion about the same axis, fig. 7.21. Thus, when the work-table has been positioned at the required height and radial position, the table can be securely clamped. The slide design ensures that the work-table surface is square to the machine spindle axis at all positions and has the advantage of ease of manufacture.

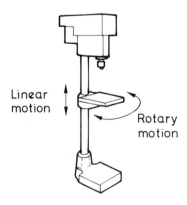

Fig. 7.21 Degrees of freedom of a pillar drilling machine

Flat slides are common on milling machines to locate and guide the saddle on the knee and the knee on the column. Flat slides give a large bearing surface, are easy and cheap to manufacture, and facilitate accurate assembly and alignment. The larger surface gives an even distribution of loads encountered during metal cutting. Adjustment for wear is provided by the use of gib strips, as shown in fig. 7.22. The gib strips are tapered along their length, and any wear which takes place between the slides is taken up by pushing in the gib strip further between the surfaces by means of an adjusting screw. The entry of swarf between the sliding surfaces is prevented by fitting wipers to the ends of the moving part – these also assist in maintaining a film of lubricant on the slide surfaces, fig. 7.23.

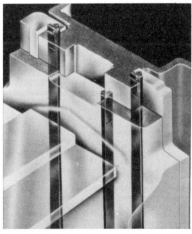

Fig. 7.22 Gib strips in a milling-machine knee

Fig. 7.23 Slide wipers

A combined vee and flat slide is common between the saddle and bed on centre lathes. This combination satisfies the kinematic design by constraining all motions other than the one required, i.e. linear motion along the bed. The advantage of this construction is that it is self-compensating for wear; i.e., as wear takes place, the saddle seats further down the vee but still retains accurate alignment. Two sets of slides are provided on a centre lathe – one for the saddle, the other for the tailstock. This arrangement, shown in fig. 7.24, prevents the wear on one set affecting the accuracy of the other. The saddle is prevented from lifting off the bed under the action of cutting forces by means of saddle strips which can be tightened when wear takes place. Due to the smaller bearing surface, wear is more rapid than with flat slides. Again, wipers are provided at each end of the saddle to prevent swarf accumulation and to maintain a film of lubricant.

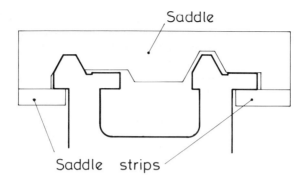

Fig. 7.24 Vee-and-flat bed of centre lathe

Dovetail slides are used when an upward movement of the slide must be prevented. These are common on milling-machine tables and on the cross and top slides of lathes, to provide movement in a straight line along one axis only. Figure 7.25 shows an arrangement of a lathe cross slide. To facilitate assembly and compensate for wear, a gib strip is fitted. This may be tapered along its length, wear being taken up by pushing the gib strip between the sliding faces by means of the adjusting screw. Alternatively, it may be parallel along its length and adjustment made by pushing the gib strip against one sliding face by means of a series of screws spaced along the length of the slide as shown on the top slide in fig. 7.25.

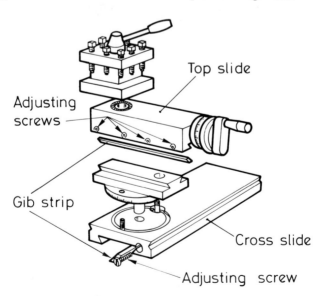

Fig. 7.25 Gib-strip arrangement of centre-lathe cross slide and top slide

Friction exists between sliding surfaces and can lead to a condition known as stick-slip. Stick-slip is the alternate sticking and slipping of a slide operated at low speed in overcoming static friction. This makes fine adjustment of slide movement difficult. For example, if the cross slide of a centre lathe is to be adjusted in order to remove 0.04 mm from the diameter of a workpiece, the slide must move 0.02 mm. If sticking of the slide occurs, no cut will take place. If a second adjustment of 0.02 mm is made, the sticking may be overcome and the cross slide may move forward the total amount of the two adjustments, i.e. 0.04 mm – thus 0.08 mm is removed from the diameter instead of the intended 0.04 mm.

To overcome the problems of stick-slip and provide easier manual operation with greater sensitivity of movement, anti-friction devices may be incorporated in slide systems. These may incorporate balls or rollers running on hardened and ground guideways to substitute rolling friction for sliding friction. An example of this principle is shown in fig. 7.26. Another method is to employ hydrostatic bearings, where oil under pressure is pumped between the sliding surfaces so that they operate on a constant film of oil.

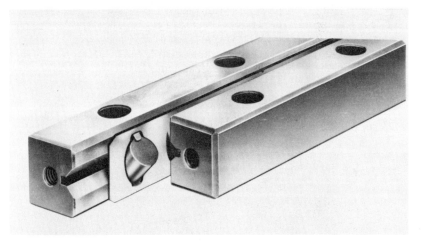

Fig. 7.26 Roller-bearing slide

Motion can be transmitted to a slide by a variety of methods, whether power-fed or hand-controlled. Hydraulic methods are employed for the power-operated reciprocating slide of a surface grinder, while a rack-and-pinion mechanism is used when the slide is hand-controlled. A series of gears and a mechanical linkage is used for power transmission to the ram of a shaping machine. The most common method for the movement of slides is by means of a screw and nut. Depending on the application, either the nut or the screw may move with the sliding member.

In the case of a cross slide of a centre lathe, for example, the screw is constrained on the saddle so that it is capable only of rotation. The nut is attached to the underside of the cross slide so that, as the screw is rotated, the nut and the slide attached to it advance along the screw, thus controlling the slide movement, fig. 7.27. In the case of a slide of a milling machine, the screw – again capable only of rotation – moves with the slide; the nut is stationary and is attached to the saddle. As the screw is rotated, it screws itself through the nut, thus controlling the movement of the slide attached to it, fig. 7.28.

Any screw and nut will have a certain amount of clearance between the threads and, as a result, any reversal of rotation of the screw will not give

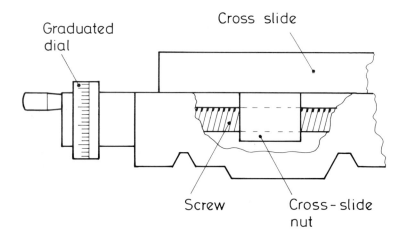

Fig. 7.27 Centre-lathe cross-slide screw and nut

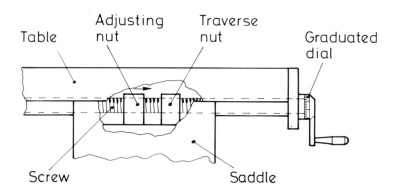

Fig. 7.28 Milling-machine-table screw and nut

slide movement until the clearance has been taken up. This clearance – known as backlash – must be allowed for by maintaining screw rotation in the same direction during operation. Backlash in a milling machine can cause problems, depending on the direction of the forces acting during cutting, and so a backlash eliminator may be fitted. This can take the form of a second nut, fig. 7.28, which is adjustable in an axial direction, this adjustment being carried out from the front of the machine.

For precise control of slide movement, graduated dials are fitted to the end of the screw. By relating the size of the dial and the number of graduations to the lead of the screw, differing degrees of accuracy can be obtained.

The lead of a screw is the distance the thread will advance in an axial direction when turned through one revolution. In the case of a single-start thread, the lead equals the pitch; for example, a cross-slide screw with a pitch of 2.5 mm will move the slide this amount for each revolution. If a calibrated dial with 250 graduations is fitted at the end of the screw, each graduation will represent a slide movement of (2.5 mm)/250 = 0.01 mm. The diameter of the dial will determine the space between adjacent graduations, which is important for ease of reading; for example, in the case above, an 80 mm diameter dial would give graduations approximately 1 mm apart (i.e. $\pi D/250$), a 120 mm diameter dial would give graduations approximately 1.5 mm apart, and so on.

Square threads are difficult to produce economically, because of their perpendicular sides, and for most machine-tool applications they have been largely replaced by the Acme thread, the form of which is shown in fig. 7.29. The efficiency of this type of screw is low, due to the effects of

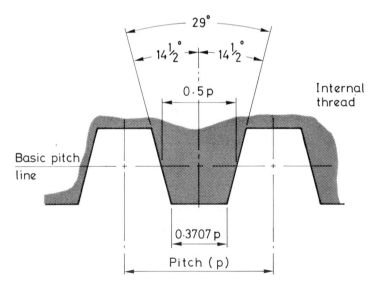

Fig. 7.29 Acme screw-thread profile

friction, and, as already stated, a certain amount of backlash exists between the screw and the nut. These disadvantages can be overcome to a great extent by the use of ball screws. The ball-screw assembly consists of a screw and a nut each containing a track in which balls are circulated, fig. 7.30. The balls are contained within the nut and, as rotation between the screw and nut takes place, the balls are deflected and led back to the start point, so giving a continuous flow during rotation. Friction is considerably reduced and thus less power is required to drive either the screw or the nut. As a result of the reduced friction, mechanical efficiency as high as 90% is possible. Backlash can be eliminated by fitting two ball nuts preloaded against each other, i.e. with the nuts pushing away from each other so that the balls bear on opposite flanks of the ball track.

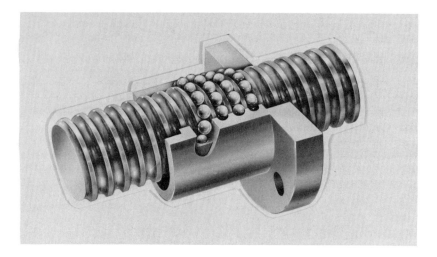

Fig. 7.30 Ball screw

7.4 Accuracy of slideway systems

Specific objective To state the standards of accuracy to be expected from slideway systems.

The accuracy of the slideway system of a machine tool will affect the accuracy of any workpiece produced on it. This involves the straightness of the slideway and its relative direction of motion with other motions, i.e. parallelism and squareness. The permissible deviations or errors are set out in the various parts of BS 4656, 'Specification for the accuracy of machine tools and methods of test'. The ways in which tests for accuracy are carried out are dealt with in section 7.5.

Drilling machines

The important considerations for a pillar drilling machine are straightness of the pillar and squareness of the spindle axis to the table surface. In both cases, the permissible deviations according to BS 4656:part 11:1974 are 0.06 mm per 300 mm length.

Milling machines

The important considerations for knee-and-column-type horizontal and vertical milling machines are straightness of the vertical movement of the knee, squareness of the table surface to the column slides for the knee, and parallelism of the table surface to its movement in both longitudinal and transverse directions. In each case, the permissible deviation according to BS 4656:part 3:1971 is 0.025 mm for any length of 300 mm. In the case of parallelism of the table surface to its movement, a maximum tolerance of 0.05 mm is specified.

Centre lathes

For general-purpose centre lathes, the important considerations are straightness of the bed slideways, straightness of the saddle movement with reference to the axis of the machine centres, parallelism of the spindle axis to the saddle longitudinal movement, and squareness of the transverse movement of the cross slide to the spindle axis.

The permissible deviations for general-purpose centre lathes are specified in BS 4656:part 1:1981 according to the size of lathe, i.e. the distance between centres and the swing.

For lathes with a centre distance greater than 500 mm but less than 1000 mm and with a swing not greater than 800 mm, the permissible deviation for straightness of bed slideways is 0.02 mm convex.

The permissible deviation for straightness of their saddle movement with reference to the axis of the machine centres is 0.02 mm.

The permissible deviation for parallelism of their spindle axis to the saddle longitudinal movement is 0.02 mm per 300 mm length upwards from the spindle nose, and that for squareness of the transverse movement of the cross slide to the spindle axis is 0.02 mm per 300 mm length.

Surface-grinding machines

In the case of surface-grinding machines with a horizontal spindle and a reciprocating table, the important considerations are straightness of the table slideways, parallelism of the table surface to its movement in both longitudinal and transverse directions, squareness of the longitudinal movement of the table to its transverse movement, and squareness and straightness of the vertical movement of the wheelhead to the table surface. The permissible deviations are specified in BS 4656:part 7:1971.

The permissible deviation for straightness of the table slideways is 0.02 mm up to 1000 mm with an additional 0.02 mm for each 1000 mm increase in length. The maximum permissible deviation is 0.05 mm.

The permissible deviation for parallelism of the table surface to its movement in a longitudinal direction is 0.015 mm up to 1000 mm with an additional 0.01 mm for each 1000 mm increase in length. The maximum permissible deviation is 0.05 mm. In a transverse direction, the permissible deviation is 0.01 mm up to 1000 mm length.

The permissible deviation for squareness of the longitudinal movement of the table to its transverse movement is 0.03 mm per 300 mm length.

The permissible deviation for squareness and straightness of the vertical movement of the wheelhead to the table surface is 0.04 mm per 300 mm length.

7.5 Alignment testing

Specific objective To use test mandrels and spirit levels for alignment testing of machine tools.

Alignment testing of machine tools is carried out to determine the accuracy, position, and relative movements of elements of a machine which may affect any workpiece produced on it.

Methods for testing the accuracy of machine tools are covered by BS 3800:1964 (1979), while the tests and accuracy requirements for particular types of machine tool are dealt with in BS 4656, 'Specification for the accuracy of machine tools and methods of test', which is published in a number of parts.

Geometrical tests are carried out by the manufacturer of the machine tool during building and subsequent inspection to comply with the British Standards. This ensures that the user can produce work from the machine tool to the desired degree of accuracy.

A user may require to carry out some geometrical tests during machine installation or on old machine tools which have been dismantled for reconditioning and subsequently rebuilt.

Geometrical tests are carried out to determine

a) straightness, e.g. of slideways;
b) flatness, e.g. of the table surface;
c) parallelism, e.g. of the spindle axis to the table surface;
d) squareness, e.g. of the spindle axis to the table surface;
e) rotation, e.g. of a spindle, or run-out.

Figure 7.31 shows the method used to check the straightness of the slideways of a centre lathe using a precision spirit level. Measurements are made at positions equally distributed along the length of the bed. Full details of geometrical tests and accuracy requirements for general-purpose lathes can be found in BS 4656:part 1:1981.

Figure 7.32 shows the method used to check flatness of a drilling-machine table using a precision spirit level. Measurements are made in a series of lines forming a grid so that the whole surface is covered. Full details of geometrical tests for column and pillar-type drilling machines can be found in BS 4656:part11:1974.

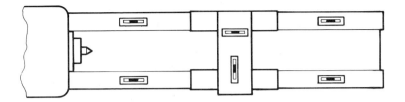

Fig. 7.31 Straightness check of centre-lathe slideways

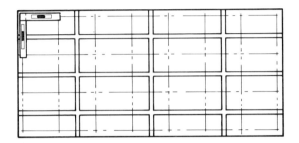

Fig. 7.32 Flatness check of a drilling-machine table

Figure 7.33 shows the method used to check the parallelism of the spindle axis to the table surface on a horizontal milling machine, using a mandrel and a dial indicator. The dial indicator is moved on the table surface, under the mandrel at a point near to the spindle nose and at another point at a specified distance from it. Full details of geometrical tests and accuracy requirements for milling machines can be found in BS 4656:part 3:1971.

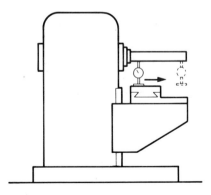

Fig. 7.33 Parallelism check of the spindle axis of a horizontal milling machine

Figure 7.34 shows the method used to check the squareness of the spindle axis to the table surface on a vertical milling machine, using a mandrel and a dial indicator. An arm carrying the dial indicator is attached to the mandrel located in the machine spindle. The stylus of the dial indicator is adjusted to touch the machine-table surface. When the spindle is revolved by hand, the dial indicator describes a circumference, the plane of which is perpendicular to the axis of rotation. Any deviation at points around the prescribed circumference will register on the dial indicator.

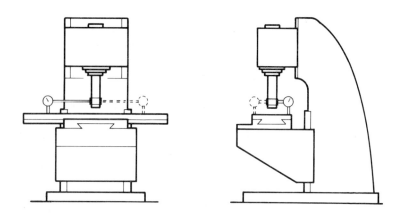

Fig. 7.34 Squareness check of the spindle axis of a vertical milling machine

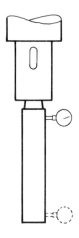

Fig. 7.35 Run-out check of the spindle taper of a drilling machine

Figure 7.35 shows the method used to check run-out of the internal taper in a drilling-machine spindle, using a mandrel and a dial indicator. The mandrel is rotated with the stylus of the dial indicator in contact, one check being made near the spindle nose and another at a specified distance from it.

Exercises on chapter 7
1 State the essential difference between a capstan lathe and a turret lathe.
2 What provision is made for the adjustment of wear of machine-tool slides, and what is the feature called?
3 What is a 'machine tool'?
4 Why is alignment testing of machine tools carried out?
5 Describe what are meant by 'forming' and 'generating' in metal cutting.
6 What is the aim of incorporating kinematic principles in machine tools?
7 Describe how a capstan lathe is more suitable than a centre lathe for large batches of work.
8 What is 'stick-slip', what effect can it have on a machine tool, and how can it be overcome?
9 Name the type of slide used when upward movement is to be prevented.
10 A cross-slide screw on a centre lathe has a pitch of 2 mm and the calibrated dial has 100 graduations. What is the value of each graduation? [*Answer*: 0.02 mm]

8 Operational planning

General objective The student describes typical machining operations and produces an operation layout of the sequence required for manufacture of given components.

The purpose of operational planning is to set out, on an operation sheet, the sequence of operations required to produce a finished component from the raw material in the most economical manner.

The way in which this is done and the amount of detail included on the operation sheet depend on the method of production used. Methods of production fall into three distinct types: job, batch, and mass production.

Job production is limited to single items or small quantities of special items and does not usually require the use of operation sheets. The way in which the items are produced is normally left to the skill of the operator, e.g. in toolroom work.

Batch production is the most common type, involving larger quantities of items produced at regular intervals, and requires the use of operation sheets.

Mass production involves the production of large quantities on a continuous basis and makes use of special-purpose machines which, once set up, run for very long periods. Operation sheets and tooling layouts are required only for the initial setting-up or for use in the design of the special-purpose machine.

Although the main purpose of operation sheets is to set out a sequence of operations, they also serve a number of other very important functions:

a) They determine the size and amount of material required. From this information, the material required can be ordered in advance and appropriate levels of material in stock can be maintained.
b) Any tooling – i.e. jigs, fixtures, gauges, etc. – required can be ordered or manufactured in advance so that it will be available at the machine when required.
c) Knowing the machines which are to be used enables machine-loading charts to be updated so that delivery dates to customers will be realistic.
d) The sequence of operations listed will enable work to be progressed through the factory in an efficient manner.
e) The inclusion of estimated times for manufacture on an operation sheet enables an accurate cost of manufacture and hence selling price to be established.

In setting out to write an operation sheet, consideration must first be given to the material from which the part is to be made. With castings and forgings, the first step is to machine a surface which can then be used as a datum for all subsequent machining operations. In machining from solid, allowance must be made for enough material to enable the part to be securely held during machining. Some simple parts can be machined completely at one setting on one machine; others require a number of separate operations on a number of different machines before final completion. Let us consider a few different types of component.

Figure 8.1 shows a fitted bolt with a hexagon head. This type of product is ideally suited to production on a capstan lathe from hexagon-bar material held in a hexagonal collet, and can be completed at a single setting. The sequence of operations is shown in operation sheet 1.

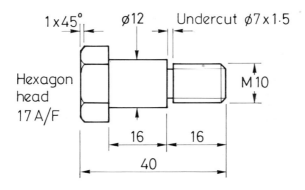

Fig. 8.1 Workpiece

Operation sheet 1
Material: 17 A/F × 42 mm hexagon bar   *Name*: Fitted bolt

Op. no.	Machine	Operation	Tooling	Position
1	Capstan lathe	Feed to stop.	Adjustable stop	Turret 1
		Turn 10 mm thread diameter.	Roller-steady box tool	Turret 2
		Turn 12 mm diameter.	Roller-steady box tool	Turret 3
		Face end and chamfer.	Face-and-chamfer tool	FTP
		Form undercut.	Undercut tool	FTP
		Die thread.	Self-opening diehead	Turret 4
		Chamfer head.	Chamfer tool	FTP
		Part off.	Parting-off tool	RTP

*FTP – front toolpost; RTP – rear toolpost

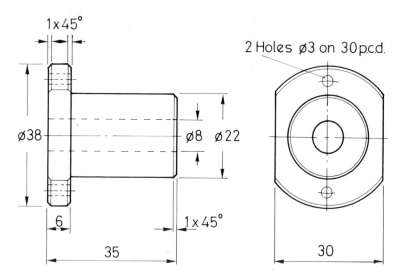

Fig. 8.2 Workpiece

Operation sheet 2
Material: ϕ 40 × 39 brass bar   *Name*: Flanged bush

Op. no.	Machine	Operation	Tooling	Position
1	Capstan lathe	Feed to stop.	Adjustable stop	Turret 1
		Turn 38 mm diameter.	Knee turning toolholder	Turret 2
		Turn 22 mm diameter and start drill.	Knee turning toolholder and centring tool	Turret 3
		Face end and form chamfers (2).	Chamfering tool	FTP
		Drill 8 mm diameter hole.	Drill ϕ 8	Turret 4
		Part off to 36 mm long.	Parting-off tool	RTP
2	Capstan lathe	Reverse and face to 35 mm length and chamfer flange diameter.	Face-and-chamfer tool	FTP
3	Horizontal milling machine	Hold in chuck mounted vertically.	Three-jaw chuck.	
		Straddle mill two flats to 30 mm dimension.	Side-and-face cutters (pair) ϕ 80; 30 mm spacing collar	
4	Bench	Remove sharp edges.		
5	Drilling machine	Load in drill jig.	Drill jig	
		Drill two holes 3 mm diameter.	Drill ϕ 3	
6	Bench	Deburr holes.		

Figure 8.2 shows a flanged bush with two flats and two holes. In this case, three types of operation are required to complete the machining, i.e. turning, milling, and drilling. The turning operation would naturally be done first and could be carried out from bar material on a capstan lathe. The decision has then to be made whether to drill the holes or to mill the flats first. It is usually easier to mill the flats first, since their radial position is not important (the holes have not yet been drilled, so there is no relative position to maintain). The part can be simply held in a chuck fixed in a vertical position and the two flats can be machined on a horizontal milling machine using a pair of side-and-face cutters (straddle milling) spaced by means of a 30 mm spacing collar.

In production work, the two holes would be drilled by loading the part into a drill jig. This provides location for the part and guides the drill into the same position each time the drilling operation is carried out. Having milled the flats first, this provides an ideal location in the jig to maintain the relative positions between flats and holes. The sequence of operations is shown in operation sheet 2.

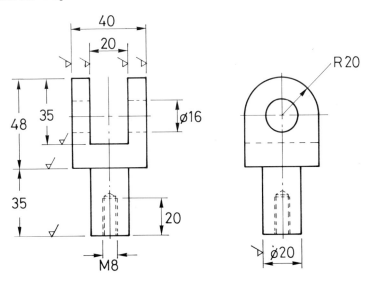

Fig. 8.3 Workpiece

Figure 8.3 shows a link to be produced from a steel forging with a machining allowance left on the surfaces indicated for machining. The simplest method of producing a datum for subsequent machining is to turn the 20 mm diameter on a capstan lathe by holding the forging in a square collet. The forging can then be held in a three-jaw chuck mounted vertically on a horizontal milling machine or in a specially designed fixture. Three side-and-face milling cutters would be used, one pair to produce the 40 mm dimension and one located between these by means of 10 mm spacing collars to produce the 20 mm wide slot. Such a set-up using three cutters is known as gang milling. Using this procedure ensures that the slot will be central to the 20 mm diameter.

The two holes can then be drilled in a drill jig, locating from the diameter and slot to ensure correct alignment. This sequence of operations is shown in operation sheet 3.

Operation sheet 3
Material: Steel forging   *Name*: link

Op. no.	Machine	Operation	Tooling	Position
1	Capstan lathe	Hold in square collet.		
		Face to length and chamfer.	Face-and-chamfer tool	FTP
		Turn 20 mm diameter to length.	Roller box tool	Turret 1
		Start drill.	Centring tool	Turret 2
		Drill 6.8 mm diameter tapping size to depth.	Drill ϕ 6.8	Turret 3
		Tap M8.		
2	Horizontal milling machine	Hold in chuck mounted vertically.	Three-jaw chuck	
		Gang mill outer faces to 40 mm and 20 mm slot.	Side-and-face cutters (pair) ϕ 160	
			Side-and-face cutter 20 mm wide ϕ 125	
			10 mm spacing collars (2)	
3	Bench	Remove sharp edges.		
4	Drilling machine	Load in drill jig.	Drill jig	
		Drill two holes 16 mm diameter.	Drill ϕ 16	
5	Bench	Deburr holes.		

Figure 8.4 shows a bearing bracket to be machined where shown from a cast-iron casting containing a cored hole. The base face requires to be ground. In this case the base face is machined first on a vertical milling machine using a shell end-milling cutter, to provide the necessary datum surface. The bracket is then held in a vice with the datum face against the fixed jaw and one end face is milled square to the datum face. The bracket is then reversed, with the datum face still located against the fixed jaw of the vice, and the previously machined end face is located on a parallel. The second end face is milled. This method ensures that the end faces are parallel to each other and square to the base face. The base face is then ground to ensure flatness, clamping the side face against an angle plate placed on the permanent-magnet chuck of a surface-grinding machine and lining up the base face.

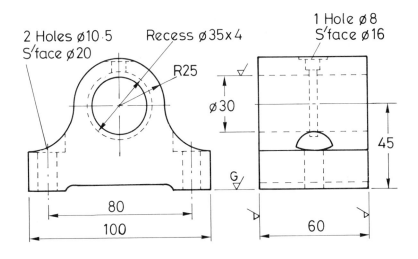

Fig. 8.4 Workpiece

Operation sheet 4
Material: Cast-iron casting   *Name*: Bearing bracket

Op. no.	Machine	Operation	Tooling
1	Vertical milling machine	Hold in vice. Machine base face. Reset in vice with base face against fixed jaw. Machine side face. Reverse and machine opposite side face to 60 mm thickness.	Shell end-milling cutter ϕ 80
2	Surface-grinding machine	Clamp side face against angle plate, lining up base face. Grind base face.	Angle plate
3	Centre lathe	Load in fixture. Bore to 30 mm diameter and form recess 35 mm diameter × 4 mm wide.	Turning fixture Boring tool Recess tool
4	Drilling machine	Load in drill jig. Drill two holes 10.5 mm and one hole 8 mm diameter. Spotface two holes 20 mm and one hole 16 mm diameter.	Drill jig Drills ϕ 8 and ϕ 10.5 Spotface cutters ϕ 16 and ϕ 20

The bore is already present in the form of a cored hole and therefore only requires boring to the correct size. This is done by locating the datum face in a fixture, mounted on the spindle of a centre lathe in place of the chuck, such that the hole in the bracket is in line with the spindle axis. The hole is bored and the lubrication recess is machined. Grinding the face before this operation ensures accuracy of dimension from the base to the bore centre.

The final operation is to drill and spotface the three holes, which would be carried out in a drill jig locating from the datum face and the bored hole.

The sequence of operations is shown in operation sheet 4.

Where short parts of constant cross-section are required, it is easier and cheaper to machine a single long length to the desired cross-section and then cut it up to produce a number of parts of the required length. By doing this, handling and machining times are reduced and the longer length is easier to hold while machining is carried out.

For example, let us consider the production of tee nuts as shown in fig. 8.5. If these parts were to be machined as single items, they would first be sawn to length plus an allowance for machining. The ends would then be machined to length and the steps be machined at each side.

When machined in a length, the bar would first be cut to a length to suit the width of the vice jaws. Let us say the vice jaws are 200 mm wide, so the length of bar can be of this length. The bar is held in the vice and the steps are machined on a horizontal milling machine by straddle milling in one pass. The vice is then turned through 90° and a narrow slitting saw is used to clean up one end of the bar. The bar is then fed out against a stop to give the required length and the slitting saw is used to cut off. The slitting operation produces the length and the finished surface to the end face in the one operation. Thus, for a length of 200 mm and using a 3 mm wide slitting saw, ten tee nuts could be produced from each single length. The sequence of operations is shown in operation sheet 5.

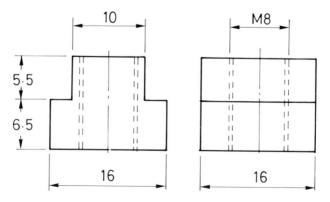

Fig. 8.5 Workpiece

Operation sheet 5

Material: Low-carbon-steel bright-drawn bar 16 × 12 × 200 (makes 10 off)
Name: Tee nut

Op. no.	Machine	Operation	Tooling
1	Horizontal milling machine	Hold in vice. Straddle mill to 10 mm width × 5.5 mm deep.	Side-and-face cutters (pair) ϕ 100
2	Horizontal milling machine	Hold in vice against end stop. Cut to 16 mm length using 3 mm wide slitting saw.	Slitting saw ϕ 100 × 3 mm wide
3	Bench	Remove sharp edges.	
4	Drilling machine	Load in drill jig. Drill and tap one hole M8.	Drill jig Drill ϕ 6.8 Tap M8
5	Bench	Deburr hole.	

Exercise on chapter 8

Compile an operation sheet for the component shown in fig. 8.6.

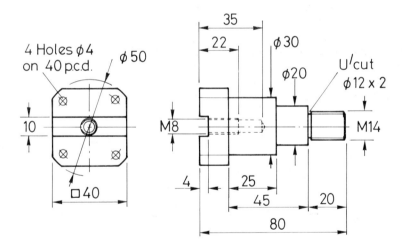

Fig. 8.6 Workpiece

Index

abrasion resistance, 135
Acme thread, 165-6
additives, 68
adjustable stop, 155
Airy points, 90
alignment testing, 168
antioxidants, 68
automatic guards, 50

backlash, 165
ballscrew, 166
balls, steel, 103
batch production, 172
bend allowance, 55
bending, 44, 55, 61-2
bend test, 20
blanking, 44, 50, 58-60
   layout, 62-6
bolster, 57
bore gauge, 118
boring-bar holder, 155
British Calibration Service, 79
British Standards Institution, 79
bulk factor, 70
burns, 23

calibration grade, 83, 88-90
capstan lathe, 152
carburising flame, 4
cemented carbide, 137-8
centre lathe, 167
centring-and-facing tool, 156
ceramic, 139
chip
   continuous, 121
   continuous with built-up edge, 123
   discontinuous, 121
clearance, 51-2
closed-die forging, 38
cold rolling, 32, 34

cold working, 37
colourants, 69
column-type power press, 47
comparative measurement, 112
compression moulding, 69-71
continuous chip, 121
continuous chip with built-up edge, 123
cope, 28-30
copying, 147-9
core, 30-1
core prints, 30-1
cracks, in welding, 20
cubic boron nitride, 139
curing, 67
cutting fluids, 140
cutting force, 123
cutting speed, 125-7
cutting tools, 120
   materials, 135-40

degrees of freedom, 159
destructive testing, 16
dial calipers
   external, 117
   internal, 118
dial indicators, 113-15
diamond, 139
die, 44, 50, 57
diehead, 158
die-sets, 57
direct extrusion, 35
discontinuous chip, 121
dovetail slide, 162
draft, 30, 38
drag, 28-30
drawing, 37
drilling machine, 150, 167
drop forging, 38-9
ductility, 27
dust, 42

edge preparation, 15
electric arc welding, 11
electric shock, 21
electrodes, 12–13
enclosed press tools, 50
extrusion, 34–7
   direct, 35
   indirect, 36

feed force, 123
feed motion, 147
feel, 113
fire, 23, 42
filler metals, 6–8
fillers, plastics, 68
fixed guards, 50
fixed stop, 58, 60
flame retardants, 60
flame, welding
   carburising, 4
   neutral, 3
   oxidising, 4
flash, 39, 71
flat centring tool, 156
flatness, 94, 168
fluidity, 27
flux, 6–8, 13
fly press, 44–5
forging, 37–9
forming, 147–9
four-high mill, 34
fumes, 24, 42

gap-frame power press, 46
gates, 30, 72
gauge blocks, 80–7
   accessories, 86–7
generating, 147–9
generators, 12
geometrical tests, 168
gib strip, 161–2
guards, 50

hazards in primary process work, 40–3
hearing protectors, 41
high-carbon steel, 136
high-speed steel, 136
hot rolling, 31–2
hydraulic power press, 48

indirect extrusion, 36
infra-red radiation, 22–3
injection moulding, 72–4
inserts, 74–5
interlocked guard, 50
International Organisation for Standardisation, 79

job production, 172

kerf, 8
kinematic design, 160
knee turning toolholder, 156

lathe
   capstan, 152
   centre, 167
   turret, 152
lead, 165
leftward welding, 5
length bars, 88
   accessories, 91
length standards, 80
light-screen guards, 50
lubricants, 68

machine slides, 160–5
machine tools, 146
malleability, 27
manual metal arc welding, 11
mass production, 172
materials handling, 42–3
material utilisation, 63
measurement
   comparative, 112
   external tapers, 110
   internal tapers, 107
   plain bores, 105
mechanical power press, 46, 48
metal-removal rate, 128
metre, 80
milling machine, 151, 167
mill, rolling,
   four-high, 34
   three-high continuous, 32–3
   two-high reversing, 32
monomers, 67
mould, 71, 73

moulding
   compression, 69–71
   injection, 72–4
   tools, 70, 74
   transfer, 71

National Physical Laboratory, 79
neat cutting oils, 141
negative rake, 132–4
neutral axis, 55
neutral flame, 3
noise, 41
non-destructive testing, 16–17

oblique cutting, 123
open-fronted power press, 46
operational planning, 172
operation sheet, 172
orthogonal cutting, 123
overlap, 20
oxidising flame, 4
oxy-acetylene cutting, 8
oxy-acetylene welding, 2

parallelism, 168
parting line, 28
parting sand, 29
pattern, 28–30
personal protection, 41
penetration, 5, 19
piercing, 44, 50, 60
plasticisers, 68
plasticity, 27
plastics moulding, 67
polymerisation, 67
porosity, 19
positive rake, 132–4
power, 128–30, 150
power consumption, 128–30
power press, 45–8
   column type, 47
   gap-frame, 46
   hydraulic, 48
   mechanical, 46, 48
   open-fronted, 46
   straight-sided, 47
Power Presses Regulations, 49
power transmission, 150
preform, 71

press tool, 50
presswork, 44
primary forming, 26
primary motion, 146
punch, 50, 57

radiation, 22
rake angle, 126–7, 132–4
   negative, 132–4
   positive, 132–4
rectifiers, 12
red hardness, 126, 135
reference grade, 88
rightward welding, 6
riser pin, 30
roller-bearing slide, 163
rollers, steel, 103
roller-steady box tool, 156
rolling, 31
rotation, 168
roundness, 98
runner, 71–2
runner pin, 30

safety
   in plastics moulding, 78
   signs and colours, 40–1
   in welding, 21
sand, 30–1
sand casting, 28
semi-synthetic fluids, 142
shaping machine, 150
shot size, 73
shrinkage, 30
side bending, 61–2
sine bar, 100
slag chipping, 23
slag inclusions, 20
slides, machine, 160–5
   dovetail, 162
   movement, 163–5
sliding stop, 58, 60
soluble oils, 141
spatter, 20
springback, 57
sprue, 72
squareness, 96, 168
squares, 96–7
square threads, 165
stabilisers, 68

standards, 79
steel balls, 103
steel rollers, 103
stellite, 137
stick-slip, 163
straightedges, 92-4
straightness, 92, 168
straight-sided power press, 47
stripping, 53-4
strip width, 63
surface defects, 19
surface-grinding machines, 152, 167
surface plate, 95
surface table, 95
synthetic fluids, 142

thermoplastics, 67
thermosetting plastics, 67
three-high continuous mill, 32-3
tool life, 125
toolmaker's flat, 95
toughness, 27, 135
transfer moulding, 71
transformer, 11
turret lathe, 152
two-high reversing mill, 32

ultra-violet absorbers, 69
ultra-violet radiation, 22-3
undercutting, 19

vee bending, 61-2

weld
  defects, 16
  joints, 14
  profile, 17-18
  size, 17-18
  types, 14
welding
  a.c., 11-12
  cracks in, 20
  d.c., 11-12
  definition of, 1
  electric arc, 11
  flames, 3-4
  leftward, 5
  manual metal arc, 11
  oxy-acetylene, 2
  positions 16
  processes 1
  rightward, 6